新景观 2008—2010
NEW LANDSCAPE ARCHITECTURE

绿茵景园（集团）公司
蓝调国际设计机构　编著

（第三卷）

四川出版集团　四川美术出版社

图书在版编目（CIP）数据

新景观. 第3卷, 2008~2010/绿茵景园（集团）公司, 蓝调国际设计机构著. -- 成都：四川美术出版社, 2011.4
ISBN 978-7-5410-4561-5

Ⅰ.①新… Ⅱ.①绿… ②蓝… Ⅲ.①景观—环境设计—作品集—中国—现代 Ⅳ.①TU-856

中国版本图书馆CIP数据核字（2011）第056663号

新 景 观（2008—2010）
XIN JING GUAN

绿茵景园（集团）公司
蓝调国际设计机构 编著

出 品 人	马晓峰
责任编辑	杜 娟
装帧设计	绿茵景园（集团）公司·蓝调国际设计机构
中英文责任较对	王 曼　席 勇
出版发行	四川出版集团·四川美术出版社
	成都市三洞桥路12号 邮政编码:610031
成品尺寸	250mm×250mm
印 张	27.5
图 片	1000幅
字 数	150千字
制 版	绿茵景园（集团）公司·蓝调国际设计机构
印 刷	四川盛图彩色印刷有限公司
版 次	2011年5月第1版
印 次	2011年5月第1次印刷
书 号	ISBN 978-7-5410-4561-5
定 价	350元

- 版权所有·翻印必究
- 本书如有缺页、破损、装订错误，请寄回印刷厂调换
- 如需购本书，请与本社邮购组联系
  地址：成都市三洞桥路12号
  电话：（028）87734382　邮政编码：610031

总 策 划	绿茵景园（集团）公司·蓝调国际设计机构	General Planner: CELEC Engineering Co., Ltd·Blue Tone International Design Organization
	绿茵新景观文化传播机构	CELEC New Landscape Culture Transmission Organization
	绿茵新景观设计机构	CELEC New Landscape Culture Design Organization
撰 稿	成都绿茵景园	Written by : Chengdu CELEC
	重庆绿茵景园	Chongqing CELEC
	北京绿茵景园	Beijing CELEC
	上海绿茵景园	Shanghai CELEC
特约顾问	曾跃栋　张 坪	Special Adviser: Zeng Yuedong, Zhang Ping
主 编	白 锐	Chief Editor: Bai Rui
副 主 编	敖 翔　谈晓琦	Deputy Editor: Ao xiang , Tan Xiaoqi
特约编辑	邹巧都	Special Editor: Zou Qiaodu
平面视觉	白 锐　阿 笠　邹巧都	Plane Visual: Bai Rui , A Li , Zou Qiaodu
英文审译	周 琳	English Editor: Zhou Lin

Preface

卷首语

　　本期绿茵新景观为前两期的延续，着重为您讲述2008-2010年间公司优秀的景观规划设计作品。意在诚心与大家相互交流学习，竭力为大家呈现一道风格迥异，趣味横生的视觉盛宴。

　　本期项目涵盖了多个地区多种类型的景观设计与施工项目，从住宅到旅游规划，到城市绿地公园，再到综合性景观设计等，您可以从中一览设计师在全新理念的引导下结合独特的视角，丰富其使用功能以满足当代生活需求，并将其与自然景观有机结合,重塑其活力,使场地具有多种发展的可能性，对不同类型景观空间的处理使之产生新的秩序。作为内含丰富的要素和符号，有其多元化的处理手法：或清新雅致，或浓墨重彩，或成熟简练，或童趣万分。在这里,不存在传统景观设计意义上的"完整、完全与完美"，取而代之的是发展、变化和自由。

　　在关注优秀作品的同时，本期特别专题还将为您展示绿茵历时两年设计并施工的青城·豪生（国际）酒店景观工程的全过程。大气的酒店、温婉的温泉、精致的高尔夫推杆练习场和光影交错的道缘休闲岛，共同谱写出整个酒店设计施工的华美乐章。

编者

　　The current issue of CELEC New Landscape is the continuation of last two issues. It emphatically relates to you the excellent landscape planning and design works of our company during 2008 to 2010. It provides a platform sincerely for you to communicate with and learn from each other and makes every effort to present you an utterly different and extremely interesting visual feast.

　　This project covers various landscape design and implementing projects in many areas, including residence, tourism planning, urban Greenland parks and synthesized landscape design, etc. You can see from it that the designers combine unique angles of view under the guidance of fresh concepts. In this way, the designers enrich the site's functions to satisfy the needs of the present life. They also combine the site with the landscape organically to reshape its vigor and enable it to have various developing possibilities. Dealing with different types of landscape space produces new order. As an element and symbol containing a lot, the landscape space has multiple dealing ways. It can be dealt in a fresh and refined way, or dealt with rich color. It can also be dealt in a mature and simple way, or dealt with childlike delight. In spite of integrity, complete and perfect in the sense of traditional landscape design, instead, there is development, change and freedom in the current issue.

　　While focusing on the outstanding designs, this issue presents you CELEC's 2-year process of designing and constructing Qingcheng Johnson International Hotel landscape project. The magnificent hotel, the comfortable springs, the exquisite golf course and comfortable Daoyuan leisure island all together compose the graceful musical movement of the hotel design and construction.

Editor

Homeland／家园

文/曾跃栋

复活节岛是世界上最偏僻的岛屿，当地居民那帕努尼人曾经创造出太平洋上最耀眼的文明之一，他们曾是富有创造力的农民、雕刻家、杰出的航海家，但却在人口过剩、资源缺少的夹缝中挣扎。这些岛屿上曾经生长着世界上最高的棕榈树，如今已荡然无存。那帕努尼人砍伐了它们，导致大量水土流失，他们也不能捕鱼，因为已没有树木供他们制作独木舟，他们将资源开发殆尽，结果酿成社会动荡，在暴乱与饥饿中，几乎没有人逃过劫难。

复活节岛的真正谜团不是岛上奇异的雕像，而是在于当地人为何在资源缺少的情况下不及时作出反应，他们的文明没能生存下来。但这对于今天的我们来说却有着特殊的意义。复活节岛的故事和岛上居民的遭遇或许值得我们反思：最近五十年，我们对地球的改变超过了前二十万年的总和。

地球用数十亿年为单位来计算时间，植物花了四十亿年才得以出现。在生物链中，植物就像教堂的尖顶，是一个完美的、活的雕塑，它们不服从地心引力，是唯一永远向上运动的自然物，它们不慌不忙地向着太阳生长，以获取滋养绿叶的能量，它们从微小的蓝藻那里，继承了捕获太阳能量的本领，它们将能量贮存起来供自身消耗，将能量转化为木质和叶片，然后分解为水、矿物质、植物物质和生命物质。所以，慢慢的，生命不可缺少的土壤开始形成。土壤进行着永不停息的各种活动，微生物在土壤里进食、挖掘、松土并改造土壤。这些生物活动制造了腐质土，地球上所有的生命都被这层肥沃的土壤联系起来。关于地球上的生命，我们都知道什么呢？我们了解多少物种，总数的十分之一？或者百分之一？我们了解将生命联系在一起的纽带吗？地球是一个奇迹，生命是一个谜。各种动物形成了，与它们的习性和特点一起顽强生存到了今天。有些动物适应了栖息地，栖息地也适应了它们，双方都获益，动物解决了饥饿问题，植物可以再次枝繁叶茂。

生命在地球上的探险旅程中，每个物种都有自己的角色，都有自己的地位。没有哪一种是无用的或是有害的，它们形成了一种平衡关系，造就我们——智人——聪明人——故事开始的地方。我们受益于地球馈赠的一份四十亿年之久的遗产，我们的存在不过二十万年，但是我们改变了世界的面貌。在最初的两万年里，尽管我们有自己的脆弱性，但我们占据了每一个栖息地，征服了其它动物此前无法征服的大片领地。在后来的十八万年之后，由于出现了更适宜的气候条件，人类开始定居下来，我们不再依靠捕猎来生存，我们选择了居住在湿润的环境，那里有大量的鱼类、猎物与植物，也就是土地、水与生命结合的地方。

即使在今天，人类还是大部份居住在各大洲的海岸线或河岸与湖岸附近。在世界各地有四分之一的人仍然生活在六千年前的原始状态，他们唯一获取的能源是大自然赋予的四季变化。有十五亿人沿用这种生活方式，超过了富裕国家人口的总和。但人类寿命短暂，大自然的不可预知加重了日常生活的负担。人类智慧在于时常洞悉自己的弱点，人类驯化并利用动物，用来扩张领土，弥补了自然本来赋予我们的体力与力量上的不足，但是如果吃不饱，我们如何去征服世界？农业的发明最终彻底改变了我们的历史。它的存在仅仅不到一万年时间，是人类第一场伟大革命，人类开始有了剩余产品，这导致了城市与文明的出现。数千年艰苦历史的记忆逐渐淡忘了，我们学会了在不同土壤和气候下进行农作物的耕种，增加了农作物的收成和种类。我们和其它物种一样，每天首要任务是解决温饱问题。当土壤不再肥沃，水资源开始匮乏时，我们会付出极大努力改造干旱土地，使其适应农作物的生长。人类以极大的忍耐力与奉献精神改变着土地，近乎祭神仪式般的不停的重复着。农业目前仍然是世界上最普通的职业，约有半数的人类仍在耕种土地，其中超过四分之三的人仍是手工操作。农业像传统般一代一代有血有汗的星火相传，因为它是人类生存的先决条件。在依赖了体力劳动很长一段时间后，人类开始发掘地球深处的能量，这些能量也来自植物，是阳光的聚合体，是数以亿计的植物在亿万年前捕获的纯能量，即太阳能，那是煤，是天然气，最重要的是石油。这种太阳光聚合体将人类从手工劳作中解放出来，能源令人类摆脱了时间枷锁，由于能源，一部份人获得了前所未有的舒适生活，在五十年里，仅仅一代人的时间，我们使地球发生了前所未有的改变。

越来越快，在过去六十年中，地球上人口倍增，超过二十亿人移居城市。

越来越快，满眼摩天大楼拥有数百万居民的中国深圳，在四十年前只不过是一个偏僻的小渔村。

land

越来越快,上海在二十年内建造了三千座高楼,还有数百座正在建设中。

今天全世界七十亿人口有半数居住在城市。

湿地占了全世界面积的百分之六,在恬静的水面下,存在着一个天然工厂,通过这种极端的丰富与多样性的结合来过滤水份并消除所有污染。这些沼泽是水再生与净化不可或缺的环境,它们如同海绵一样调节水流,雨季吸收水份,旱季将水份释放出来。

我们为了开发更多的土地,往往把沼泽变成牧场、耕地或用于建造房屋。在上个世纪,全球半数的湿地都被抽干,我们竟不知它的富饶和所起到的巨大作用。所有生命都是相互关联的,水、空气、土壤、植物于我们就像一场无边际的魔术表演:植物将地下水以薄雾的形式释放到大气中,它们构成一个缓解暴雨侵蚀土壤的保护层。森林提供生命必须的湿润环境,它们贮存碳,它们拥有的碳超过大气中的所有含量,它们是我们赖以生存的气候平衡基石。植物提供了地球四分之三的生存环境,也就是所有生命的大部分栖息地。森林提供了我们各种补救措施,植物分泌的物质能被我们的身体识别,我们的细胞与之使用同一语言,我们属于同一家族。但是在最近的四十年里,世界最大的热带雨林亚马逊,其面积已经缩小了百分之二十。源于人类的眼前的急需和自私,森林让步给种植业和工业。世界第四大岛屿佳里曼丹岛在二十年前还覆盖着广阔的原始森林,以现在的森林破坏速度,这些森林将在十年内完全消失。生命物质将水、泥土和阳光联系在一起,在佳里曼丹作为地球生物多样性保留地之一,这种联系已被破坏了。树木砍伐殆尽后,土壤无法保留在原地,雨水将土壤从山坡上冲刷下来,甚至直接冲入大海。雨水的侵蚀使土壤更加贫瘠,千万年才形成的薄薄土层因为丧失了植被而消失了。人类只有不超过十年的时间来逆转这个趋势,

以避免让包括人类在内的地球生物进入一个我们从不了解的生存状态。我们曾经按照自己的想像改变了地球,如果我们不对我们造成的一切后果负责,那么我们如何能够承受本世纪末九十亿人口的沉重负担。

世界工业化以来,我们行为的代价是非常昂贵的。很多人并没有积极参与但却不得不付出代价。我见过在非洲沙漠上延绵的难民营大得像个城市,那种侵入心灵深处的震撼与悲伤我无法用语言来陈述。

但是我们知道只有悲观是不够的,世界上大多数国家的政府已经行动起来,保护地球上百分之二的水源。虽然不多,但已经是十年前的两倍,第一批自然公园的保护始于一百年前,它们占大陆面积的百分之十三,它们创造了人类活动与保护物种、土壤与地貌相互和谐的空间,人类与自然的共处成为定律。

在美国,纽约已认识到了自然对人类的作用,森林与湖泊提供了该市所需的饮用水。

在韩国,森林遭到战争破坏,但由于政府的造林计划,耗时五十年,森林再次覆盖了全国国土面积的百分之六十五。

超过百分之七十五的纸张是能够循环使用的。哥斯达黎加在军队与保护环境两者之间作出了选择,该国不再拥有军队,它选择了将资源贡献给了教育、生态旅游与原始森林的保护。

加蓬是世界上最大的木材生产国。它强制执行选择性砍伐,每公顷伐木数量不超过一株。它的森林是这个国家最重要的经济资源,这些森林如今有时间获得再生,现在已有保证可持续的森林管理计划。

我见过海南岛的渔民在意捕捞对象远甚于保护海洋资源;我见过德国弗莱堡一个五千人生活在其中的世界第一批生态能源社区;我见过在1998年长江特大洪水后,江泽民主席对全中国发

出的有效的福荫子孙后代的"禁伐令";我见过中国许多的城市的开发者、建设者们仅仅用了不到五年时间,以景观的名义,用绿色、用植物大幅度地改善了我们的居住空间。在众多人诟病房价、指责开发者打造或恢复或再造生态环境是功利驱使的同时,我们应该思考,伴随近六十年来的破坏,今天,我们每一个人的居住环境是不是已经或开始有了改善,若是,我们更多的应该是褒奖。

是该我们团结一起的时候了。重要的不是我们失去了什么,而是我们现在还拥有什么。我们仍然拥有半个世界的森林,数以千计的河流、湖泊和冰川,以及数以千计的生物物种,我们有力量去改变世界,那我们还等什么呢?

Homeland

Zeng Yuedong

Easter Island is the remotest island in the world. The local residents Rapa Nui have once created one of the most shining civilizations on the Pacific Ocean. There were once the most creative farmers, sculptors and excellent navigators on the island. However, they were struggling against overpopulation and lack of resources. There were once the tallest palm trees growing on the island. But now they have disappeared. That Rapa Nui cut down the trees caused severe soil erosion. They could no longer go fishing because there were no trees for them to make canoes. They have used up the resources and the lack of resource led to social unrest. In the riot and hunger, almost no one had escaped unscathed.

The real mystery of Easter Island does not lie in the amazing sculpture, but in why local people did not respond to the problem in time when the recourses were in reduction. That their civilization has not survived means a lot to us today. The story of Easter Island and the suffering of the residents may be worth our reflection. The changes we have made for the earth in recent fifty years are more than what we have done in the early two hundred thousand years.

With a billion years as a counting unit for the earth, it cost four million years for plants to come into being. In the ecological food chain, the plant, like the church spire, is a perfect and living sculpture. It does not obey the gravity and is the only natural object that is growing upward for ever and ever. Plants grow upw-ard towards the sun in no hurry to gain energy used to nourish leaves. They inherit the skill of gaining sun energy from the tiny algae. They store energy for their own consumption and transfer energy into woodiness and blades that are then decomposed into water, mine-ral, plant material and life material. Therefore, the soil Which is indispensable for life gradually comes into being. The soil carries on various unceasing activities. The microbes eat in the soil and dug, loosen and rem-ake the soil. The activities of those living things prod-uce humus soil, and all the life on the earth is connec-ted by this layer of fertile soil. What do we know about the life on earth? How many species do we know? One tenth of the total? Or one a hundredth? Do we know the link that connects life? Earth is a miracle and life is a mystery. All kinds of animals appeared on the earth and have survived till now with their habits and char-acters. Some animals adapt to their habitats and the habitats also adapt to them. Both of them benefit from each other. The animals solve the problem of hunger and plants can grow lush again.

In the expedition of life on earth, each species has its own character and position. There is no useless or harmful species. All the species form a balanced relatio-nship and make a place where the story begins for us who are wise and clever. We benefit from the heritage inherited from the earth lasting four billion years long. We have existed on the earth for only two billion years, but we have changed the earth's appearance. In the early twenty thousand years, although we had vulner-ability, we took up all the habitats and conquered the vast territory that other animals could not conquer. In the following one hundred and eighty thousand years, human began to settle down owing to a more favorable climate condition. We did not rely on hunting any more and chose to live in mild environment where there was plenty of fish, prey and plants. That is the place where land and water connect with life.

Even today, most people live along the coastlines of the continents or near the banks of rivers and lakes. One fourth of the people in the world still live in a pri-mitive condition the same as six thousand years ago. They gain energy only from season changes of nature. There are 1.5 billion people living in this way, the num-ber of which is more than the total number of people in rich countries. However, human beings have a short life span, and the unpredictable nature increases the burden of everyday life. Human's wisdom lies in their insight into their weakness. People domesticate and use anim-als to expand their territory. In this way, people make up to their weakness in body and strength endowed by nature. However, if we haven't eaten our fill, how can we conquer the world? The invention of agriculture has finally changed our history. It is human's first revolu-tion and has existed for less than ten thousand years. People begin to have surplus products, resulting in the occurrence of cities and civilization. The memory of thousands of years of arduous struggle history is grad-ually forgotten. We have leant to cultivate crops under different conditions of soil and climate and to increase the growth and kinds of crops. Just like other species, our primary task in a day is to dress warmly and eat our fill. When soil is no longer fertile and water resource begins to be in severe shortage, we will take great effo-rts to reform arid land to make it suitable for crops to grow. People are changing the land with great enduran-ce and dedication, and are repeating the changing near-ly like a libation ceremony. So far, agriculture is still the most ordinary career in the world and almost half of people are still cultivating land, more than three quarte-rs of which are relying on manual operation. Agricultu-re passes on from one generation to another with sweat like a tradition, because it is the prerequisite for human to survive. After relying on manual labor for a long per-iod of time, people begin to explore the energy deep in the earth. The energy also comes from plants and is the combination of sunshine. It is solar energy that is the pure energy gained by billions of plants in billions of years ago. The energy can be released by coal and gas and the most important one, oil. The solar combination releases people from manual labor. Energy enables peo-ple to get rid of shackles of time and provides part of people with the comfort which never existed before. In fifty years, just a period of one generation, we have ch-anged the earth to the greatest degree.

With a faster speed, the population has increased

l a n d

in the past sixty years. More than two billion people moved to cities.

With a faster speed, Shenzhen in China which was a remote fishing village forty years ago has been changed into a city full of skyscrapers.

With a faster speed, three thousand high buildings are built in twenty years. There are still hundreds of buildings in construction.

Today, half of the seven billion people in the world are living in cities.

Wetland covers six percent of the world area. Under the tranquil water, there is a factory that filters water and eliminates all pollution through extreme combination of variety and multiplicity. Those marshes are indispensable environment for water s regeneration. They control the current like a sponge to absorb water in dry seasons and to release water in rainy seasons.

In order to develop more land, the marshes are usually changed into pastures, cultivated land and constructing land. In last century, half of the wetland on the earth has been dried and we unexpectedly do not know its fertility and the important role it plays. All the life is connected. Water, air, soil and plants are like a magic show for us. Plants release the water under the ground to the air with a form of mist which constitutes a protecting layer to ease the soil erosion by rainstorms. Forest provides life with necessary wild environment and stores carbon the amount of which is more than that in the atmosphere. Forest is the balance stone for our survival. Plants provide the earth three quarters of surviving environment which is mostly the habitat for all the living things. Forest offers us various remedial measures. The material that plants secrete can be identified by our body, because our cells and theirs share the same language and we belong to the same family. However, in recent forty years, the area of the biggest tropical forest, Amazon, has been reduced by twenty percent. Owing to people' s current need and selfishness, forests give way to planting and industry. The fourth biggest island

Kalimantan Island was covered with vast primary forest twenty years ago. However, the forests will disappear in ten years with such a reduction speed. Life material connects water, soil and sunshine. The connection on Kalimantan Island which is one of biological diversity reservations has been broken. When trees are cut out, the soil is washed down from the slope and even to the sea. The erosion of rain makes soil barren and the thin soil layer formed in thousands of years disappears because of losing the vegetation.

Human has only less than ten years to reverse this trend to avoid species on earth including human entering an existing condition we don't know. We have changed the earth with our own imagination. If we don't take responsibility for what we have done, we will not be able to endure the heavy burden of nine billion people at the end of this century.

Since the industrialization of the world, the cost of our behavior is very expensive. Many people haven't participated, but have to pay the price. I have seen the refugee camp that is as big as a city spreading in the desert. I can't express by word the shock and sadness which invade the deep part in my heart.

However, it is not enough for us to be pessimistic. The governments of most of the countries have already taken actions to protect two percent of water resource on the earth. Although it is not plenty, it is twice as much as ten years ago. The protection of the first batch of natural gardens began one hundred years ago. The gardens cover thirteen percent of the continent area and have created harmonious space for human activities and species protection, soil and the landscape. The co-existence of human and nature has become a law.

In America, the effect of nature on human has been realized in New York. Forests and lakes provide drinking water for the city.

In Korea, forests were destructed by wars. However, owing to Afforestation Plan of the government taking five years, forests again cover sixty-five percent of the area in the country.

More than seventy-five percent of paper can be recycled. Costa Rica made a choice between the army and environment protection. There was no longer any troop. The resources were contributed to education, ecology, tourism and the protection of primitive forests.

Gabonese is the biggest wood producer country. It enforces selective cutting. Less than one tree is cut per hectare. Its forest is the most important economy resource for the country. Now there is time for the forest to be regenerated and there is a forest management plan to ensure sustainable development.

I have seen that fishermen in Hainan Island care about fishing objects more than sea resource protection. I have seen the first batch of ecological resource community in Freiburg in Germany where five thousand people live. I have seen that after the catastrophic flood in Yangtze River in 1998, Mr. Jiang Zemin issued an Exploitation Ban which can protect our descendants effectively. I have seen that in many cities in China, developers and constructors spend less than five years improving our living space to a great degree with green in the name of landscape. Many people complain about the high house price and criticize the developers that they are driven by utilitarianism when they recreate, restore or rebuilding ecological environment. At the same time, we should think about the destruction in recent sixty years and consider whether our living environment has been improved. If it is, we deserve praise.

It is time for us to unite. What we have lost is not important, while what we can own is important. We still own forests covering half of the world, thousands of rivers, lakes and glaciers, and thousands of species. We have the power to change the world. Then what on earth are we waiting for?

About CELEC

公司简介
www.chinacelec.com

成都绿茵景园工程有限公司作为中国境内专业从事环境景观工程设计与施工的企业，以卓越的专业品质取得了风景园林设计乙级和国家二级城市园林绿化资质，入选园林绿化协会会员单位，《中国园林》、《景观设计》的理事单位，多年蝉联最佳园林景观企业，2008年跻身于中国景观建筑100强企业之列，已发展成为中国一流的景观设计、施工营造商。

1998年，绿茵景园开始创业历程，这个充满无限生机和活力的团队经过十多年的蓬勃发展，先后在成都、北京、重庆、上海成立四家公司，业绩遍布四川、贵州、云南、陕西、山东、山西、安徽、福建、新疆、北京、重庆、上海、天津等省市，现已在国内完成各类大中型设计施工项目1000余项，设计年产值超过6000万元，施工年产值超过25000万元，由绿茵景园设计和施工的项目精品佳作不断且在业界好评如潮。

绿茵景园以"诚信为本，励精图治"为企业理念，全方位打造高效的企业运营机制和规范的企业管理制度，率先通过国家 ISO9001:2000 质量认证。与此同时，绿茵景园秉承"海纳百川"的包容精神，不断学习国内外先进的设计理念和工程技术，吸纳优秀的管理和专业人才，依靠精尖的专业设计团队、施工队伍与设备，为社会和大众创造美好的生活环境，为客户提供精益求精的产品和至诚至微的服务。

作为一个有理想、负责任的环境景观营造商，绿茵景园在行业内率先发起了为之瞩目的"新景观文化"建设并出版了一系列《新景观》书刊，引领行业新风尚，不断创新与突破，为环境景观的可持续发展出谋划策。

绿茵景园愿意一如既往地为各类客户提供专业优质的服务，为环境景观建设贡献我们的专长。

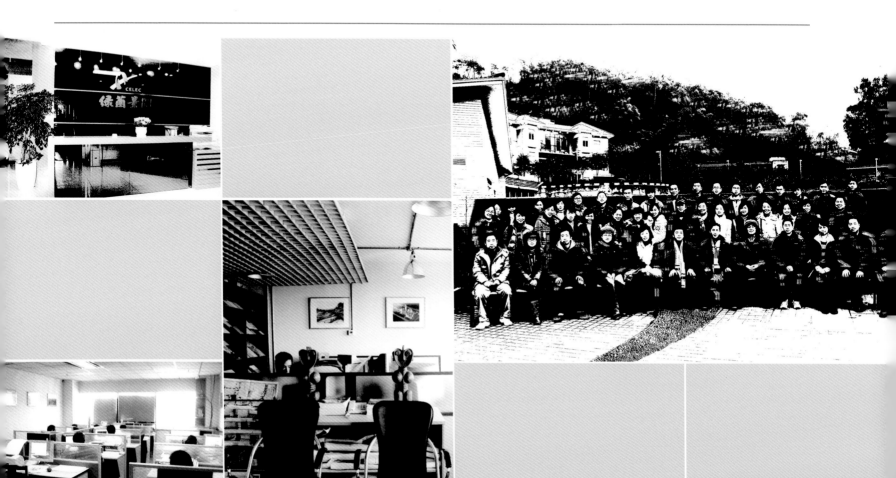

About CELEC

As a professional enterprise engaging in environmental landscape planning and project implementation in China, Chengdu CELEC has gained itself qualifications of second-class landscape designing and national second-rate urban landscaping. Enjoying its membership of landscaping society, CELEC is also a director unit of Journal of Chinese Landscaping and Journal of Landscaping. For years, it won continuously the Best Landscaping Enterprise. In 2008, CELEC upgraded itself as one of the Landscaping 100 in China. It has been the first class company of landscape planning and project implementation.

In 1998, CELEC Engineering Co., Ltd embarked on its journey of career. Having coming through rapid development, companies were in turn set up in Chengdu, Beijing, Chongqing, and Shanghai. Projects in the charge of CELEC were in massive regions like Sichuan, Guizhou, Yunnan, Shanxi, Shandong, Shanxi, Anhui, Fujian, Xinjiang, Beijing, Chongqing, Shanghai, Tianjin etc. By now, over 1000 large-scale projects have been designed and implemented domestically, and CELEC's annual designing output has been over RMB 60 million and annual implementation output has been over RMB 250 million. It receives high praise and enjoys tremendous fame for its design and project implementation.

" Based on creditability, making great efforts for prosperity" is our corporate spirit. Focusing on culture formation and taking human-oriented spirit as our direction of serving public, CELEC led the way of getting the national ISO9001: 2000 quality authentication. Trying to form a more efficient and flexible corporate operation mechanism, which will fully upgrade its operation ability, and provide customers with better design, construction, and service, CELEC has also conducted further researches and promoted innovation in respects of corporate culture, strategy, diversity strategy, administration process, construction quality and design spirit.

As an ambitious and responsible landscape constructor, CELEC initiated the new landscape culture construction, which drew widely attention of the industry, and sponsored the New Landscape as well as relevant periodicals, which have successfully advocated a new trend of the industry. On the way of innovation and contributing the sustainable development of landscaping, CELEC never stop its endeavors.

As always, CELEC is delighted to provide all the clients with professional and quality service, and contribute our expertise to the landscaping.

别墅景观设计
Villa Landscape Design

16	保利·石象湖别墅区
24	保利·公园198·拉斐庄园
36	龙湖·长桥郡
40	龙湖·江与城·原山
46	海航·香颂湖国际社区
56	复地·别院
64	富力城·维多利亚庄园
68	橘郡·米哈斯小镇
74	大众集团·湖滨花园

花园洋房景观设计
Western-style House Landscape Design

82	龙湖·弗莱明戈
90	龙湖·三千城
94	戛纳印象
102	蜀山栖镇
108	维丰·蓝湖熙岸

高层景观设计
High-rise Landscape Design

116	保利·公园198·丁香郡
124	保利·云山国际
130	龙湖·三千里
134	华润·橡树湾
136	倍特·领尚
140	金科·黄金海岸
144	融汇·二期
148	天立·水晶城
154	同盛·南桥

综合性景观设计
Synthesized Landscape Design

162	天津泰达·上青城
166	富临·桃花岛
174	仁恒置地广场

城市公园绿地景观设计
Urban Park Greenland Landscape Design

182	汶川·水磨镇5.12灾后重建景观工程
188	青城山世界自然遗产灾后生态景观恢复工程
196	重庆市茶园新区假日公园
200	重庆市江津区滨江新城公园
206	重庆市江津区琅山大道公园
212	重庆市九龙坡区西彭组团J分区休闲公园

旅游区景观设计
Tourism Area Landscape Design

222	中铁二局·花水湾度假小镇
230	保利·石象湖入口景观
236	长白山国际旅游度假区南区

特别专题——酒店景观设计
Special Column Landscape Design Hotels

244	青城（豪生）国际酒店

别墅景观设计
Villa Landscape Design

景观最大化的极致探讨

别墅设计与其他住宅类型相比，最大的特点在于园林环境已成为其产品中最大的亮点，并在市场中具有最强的竞争力。对风格特色的营造已不再是传统的单一模仿，而是融合了现代人居理念、生态造景手法的景观环境最大化的极致探讨。

注重景观本身与原生地貌的结合

《园冶·兴造论》中："因者：随基势高下，体形之端正，碍木删桠，泉流石注，互相借姿……"，即因不同的地点和环境条件灵活地组景，充分攫取自然的美景为我所用。因此优秀的别墅景观应是与原始地形地貌融为一体，力求达到虽由人作，宛自天开的最高境界。

别墅景观设计
Villa Landscape Design

页码	项目
16	保利·石象别墅区 Poly,Stone Elephant Lake Villa
24	保利·公园198·拉斐庄园 Poly,198 Park,Lafayette Manor
36	龙湖·长桥郡 Longhu,long Bridge County
40	龙湖·江与城·原山 Longhu,River and City,the Original Mountain
46	海航·香颂湖国际社区 Sailing,chanson Lake Internationai Community
56	复地·别院 Fudi-House by the Park
64	富力城·维多利亚庄园 Fuli city,Victoria Manor
68	橘郡·米哈斯小镇 Orange County,Mijas Town
74	大众集团·湖滨花园 Volkswagen Group,Garden with Lakes

一期别墅区景观布局平面图
Landscape Layout Plan of First-Stage Villa Region

回归自然
景区中的第二养生居所
保利·石象湖别墅区

Returning to Nature,
the Second Health-care Dwelling Area

Poly, Stone Elephant Lake Villa

开发商：保利(成都)石象湖旅游发展有限公司
地点：成雅高速公路66公路出口、国家级生态示范区成都市蒲江县石象湖景区内
设计面积：34000平米
景观设计：成都绿茵景园 设计二所
方案主创：阿笠
参与设计师：刘鸠鸠、张洪源、邹巧都
施工图设计师：马卉、董玲、温利梅、粟凡粒、段倩
设计时间：2009年景观设计
完成阶段：施工进行中

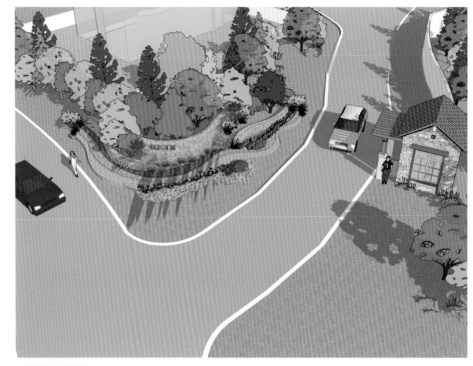
■ 别墅区入口效果图 Effect Drawing of the Entrance to Villa Region

■ 遮掩在林中的别墅区入口门卫室
Guard Room at the Entrance of Villa Region Hidden in the Forest

主题定位 Theme Orientation

奢华景区营造的"度假氛围、度假住宅"，"第二养生居所、度假气氛"

营造"生态、风情、享受的意境"，体现山中隐镇的生活气息、聚集群落、生活村的居住氛围，展示静谧的街道、山林叠院式的居住文化。

大环境的营造 The Construction of the Environment

体现半山与湖湾的区域特质：引林入院，拓居于外；林在居中，居在花湖。
道路与公共绿化空间：多用提升品质的绿化，具有第一眼震撼；不求繁，求精；不求杂，求净。
乐居空间：户户私密，家家独享，错落有致；私家空间与生态环境的互融、互衬。

小景致的塑造 Shaping of little scenes

障碍视线拓展：大放小收，错落叠加；相互借景；公共，私家。
私家领地：独享，静谧，生态健康；叠院，户外的客厅。
入户打造：延续建筑风格，形成差异化；亲人绿化，趣味装饰小景。
风情小景：墙面立体绿化，点景小品；归家小道，矮墙。

别墅景观设计 | 保利·石象湖别墅区
VILLA LANDSCAPE DESIGN

■ 一期别墅区入口设计方案　Design of the Entrance to First-Stage Villa Region

■ 别墅区公共绿化景观　Public Green Landscape of Villa Region

■ 一期入口外围实景　Actual Landscape outside the Entrance to First-Stage

Theme Orientation

The design aims at constructing resort houses with holiday atmosphere and the second health preserving place with restful atmosphere in luxurious scenic spot.

The ecological environment, amorous feelings and enjoyable artistic conception present the life flavor and living atmosphere of a secluded mountain town, and show quite streets and the living culture of mountain courtyards.

The Construction of the Environment

It Reflects the Area Features of Semi-mountain and Lakes: The construction brings the forest into yards and enables people to live into open air. The woods are in the residence and the residence is in the sea of flowers.

Roads and Public Greening Space: High-quality with impressive greening; simplicity in spite of complicity; purity in spite of mixture.

The Living Space: Every family enjoys its private and exclusive space which combines with natural surroundings and ecological environment. They add color to each other.

Shaping of Little Scenes

Extension of Obstacle Sight: To narrow and enlarge at random.

View Share: Private and public.

Private Space: Exclusive, quiet, ecologically health; folding courtyard, outdoor sitting room.

Entrance Area Design: To continue the architectural style, formin differentiation, comfortable greening and interesting adornment.

Delightful Little Scenes: Vertical greening on the wall, essays as ornament, trails leading home, parapets.

■ 别墅样板间后花园台地　Back Garden Platform of Sample Villa

■ 别墅样板间前花园广场　Front Garden Square of Sample Villa

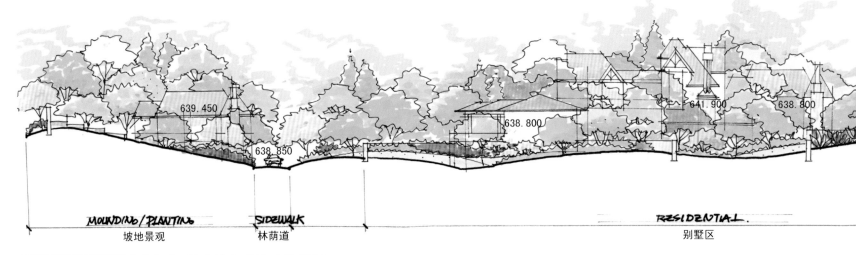

■ 别墅区建筑关系景观剖面图 Profile Drawing of the Architecture Relation in Villa Region

■ B户型别墅样板间入户平面方案 Plan Design of Type B Sample Villa

→ 设计突破重点 Breakthrough Point of the Design

① 道路景观 Road Landscape

1. 车行道：遵循地势，并营造出平面曲折与立体起伏的效果，增加趣味的同时增加景观性；
2. 人行道：林荫下的体验道漫步。

1. Driveway: It is designed to follow the terrain and creates a winding effect in the surface and an undulate effect in three dimensions. Besides, it is also added into pleasure and sense of landscape. zone and the golf green landscape, a better line of sight and the broad landscape.

2. Sidewalk: A road to experience the walk in a boulevard.

② 植物景观 Plants Landscape

1. 林荫道：减弱道路对人的干扰，优化入户的私密性；
2. 公共绿化区的统筹打造，利于整体效果展示；
3. 结合地形高差，组合植物高低搭配，预留视线通廊。

1. Boulevard: It weakens the disturbance of the road to people and optimizes the privacy for residents.
2. Overall creating of public green area is beneficial to the presentation of integral effect.
3. The design combines with the terrain difference and plants height difference, and allows a view corridor.

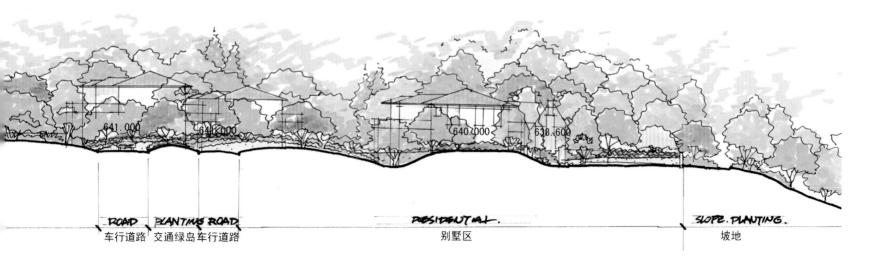

ROAD	PLANTING ROAD	RESIDENTIAL	SLOPE PLANTING
车行道路	交通绿岛 车行道路	别墅区	坡地

■ 与别墅融为一体的台地后花园　Terraced Back Garden in Combination with the Villa

■ 别墅台地后花园　Terraced Back Garden in the Villa

■ 与原生态林地交融于一体的后花园　Back Garden in Combination with Ecological Forest Land

■ 原生林地环绕着的惬意休息区
Leisurely Rest Area Surrounded by Ecological Forest Land

■ 原生林地簇拥着的无边游泳池
Boundless Swimming Pool Surrounded by Ecological Forest Land

■ 隐藏于原生林中的栈道平台
Plank Road Platform Hidden in the Ecological Forest

③ 节点景观 Joints Landscape

1. 入户景观：营造具有区分的标准入户风格，亲近感受居家氛围的第一空间；
2. 入口景观，交通转盘，点式水景，开敞绿地装饰；
3. 装饰类情景小品的引入。

1. Household Landscape: creating a standard household style with differences and the first space for residents to feel closely to household atmosphere.
2. Entrance Landscape: Traffic turntable, dot-type waterscape and open green land decoration.
3. Introduction to scenic decoration sketches.

■ 树荫下的休憩平台　　　　　　　■ 掩映在林地中的别墅一角
Rest Platform in the Shade of Trees　　A Corner of the Villa amongst the Forest

■ 原生植被衬托下的欧式雕塑
European-style Sculpture against the Background of Primary Vegetation

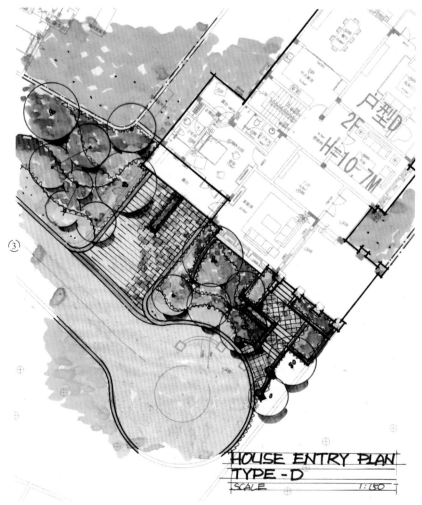

■ D户型别墅样板间入口设计平面
Entrance Design Plan of Type D Villa Sample

保利·公园198·拉斐庄园
Poly, 198 Park, Lafayette Manor

发展商：保利成都市新都区投资有限公司
项目地址：成都市新都区
设计面积：约120万平米
规划设计：美国道林建筑规划设计公司
景观方案设计：美国SWA景观设计公司
高尔夫设计公司：史密特-科里高尔夫设计公司
景观施工图设计：成都绿茵景园 设计二所
主创团队：温利梅、刘丽红、袁怡、董玲、潘宇、陈浩然
设计时间：2009-2010年

■ 总平面景观规划图 General Plan Design of Landscape

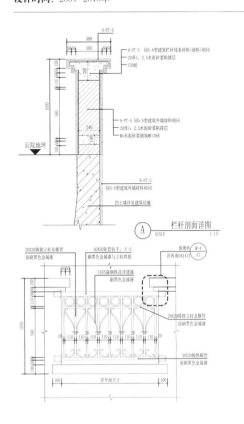

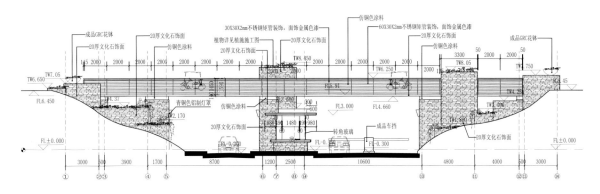

A 1-14立面图 SCALE 1:150

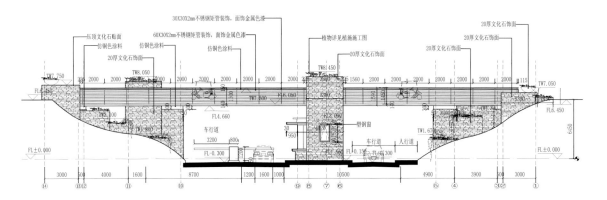

■ 别墅入口环岛景观
Circle Island Landscape at the Entrance of Villa Region

→ 设计主题 Design Theme

保利·公园198三期高尔夫社区是占地6500亩的保利·公园198中最为璀璨的一颗明珠,是主城区唯一纯高尔夫社区。项目总占地3000亩,位于成都北三环外蜀龙大道西侧,距成都市中心约15公里,距新都新城区6公里,紧邻北湖风景区和著名的熊猫基地。其中建设用地约为20%,高尔夫球场占地约2400亩。球场为国际锦标赛级18洞72杆球场,球场集锦标赛标准的线路规划,设计的无数湖泊,自然的小溪和沙坑集为一体,创造出适合不同球手的球场。

整个规划设计以美国著名的黑鹰山庄为原型,还原坡地、浅丘的自然地貌,建筑规划采用传统的欧陆风格,纯正的欧陆风格建筑呈带状分布,镶嵌于3000亩的国际锦标赛级18洞72杆高尔夫球场之中。将整个社区划分为9-10个不同地区风格的社区组团,建筑风格以英式、法式及意大利式为主。

■ 人行天桥 Pedestrian Over-bridge

■ 辽阔的天空、大面积的草坪及隐藏在植物中的建筑勾勒出美丽的画卷　A beautiful picture is formed by the spreading sky, broad grassland and the architecture hidden in the plants.

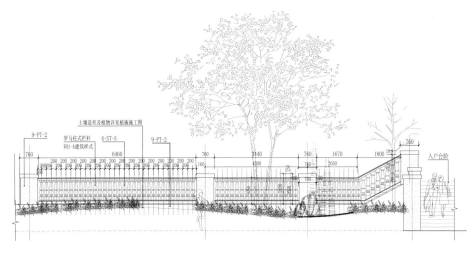

1-A型前院柱式栏杆立面图

■ 蜿蜒的车行道，中央绿化隔离带，两旁高大的乔木、花灌、草坪，打造出一条绿意昂然的归家路
A green road leading home is created by the winding driveway, central greening belt, grassland with flowers at both sides and tall trees.

■ 意大利古典式建筑入户景观
Household Landscape with Italian Classical Architecture

Design Theme

The third-stage Poly, 198 Park Golfing Community is the most beautiful and shining one in the Poly 198 Park which covers 6500 mu. It is the only golf community in the main city and the project occupies an area of 3,000 mu. It is located in the west side of Shulong Road outside the north third ring of Chengdu, about 15 kilometers away from the center of Chengdu City and 6 kilometers away from new area of Xindu City next to Beihu scenic spot and the famous panda base. Construction land covers 20% of the land and the Golf Course covers an area of about 2400 mu. The court is of International ournament standard with 18 holes and 72 poles. A course for different golfers is created by the route planning with International Tournament standard, innumerable lakes and the combination of natural streams with sand pits.

The overall design is based on the famous Black Hawk Villa in America and it restores natural landscape of hillside fields and hillocks. Besides, it adopts the traditional European continental architectural style. The community is scattered in the Golf Court of International Tournament standard with 18 holes and 72 poles covering 3,000 mu. The whole community is divided into 9-10 community groups of different styles. The main architectural is mainly in French and Italian styles.

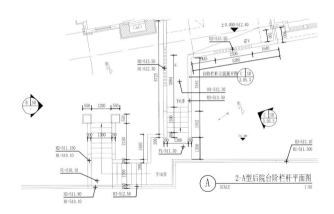

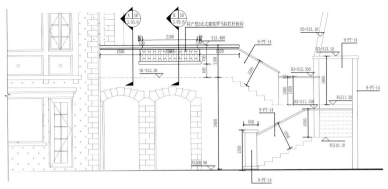

2-A型后院台阶栏杆立面图

■ 路两旁满满的鲜花，指引你回家
The road with abundant flowers at both sides leads you home

■ 转角边的楼道连接着前后院落　The corridor at the corner links the front and back yards

■ 植物中生长出的房子　The House Amongst the Plants

■ 简洁干净的庭院平台与建筑融为一体　Simple and clean yard platform integrates with the architecture

■ 花香、草香、泥土香让你感受家的温馨　　The sweet smell of flowers, grass and the soil enables you to feel at home

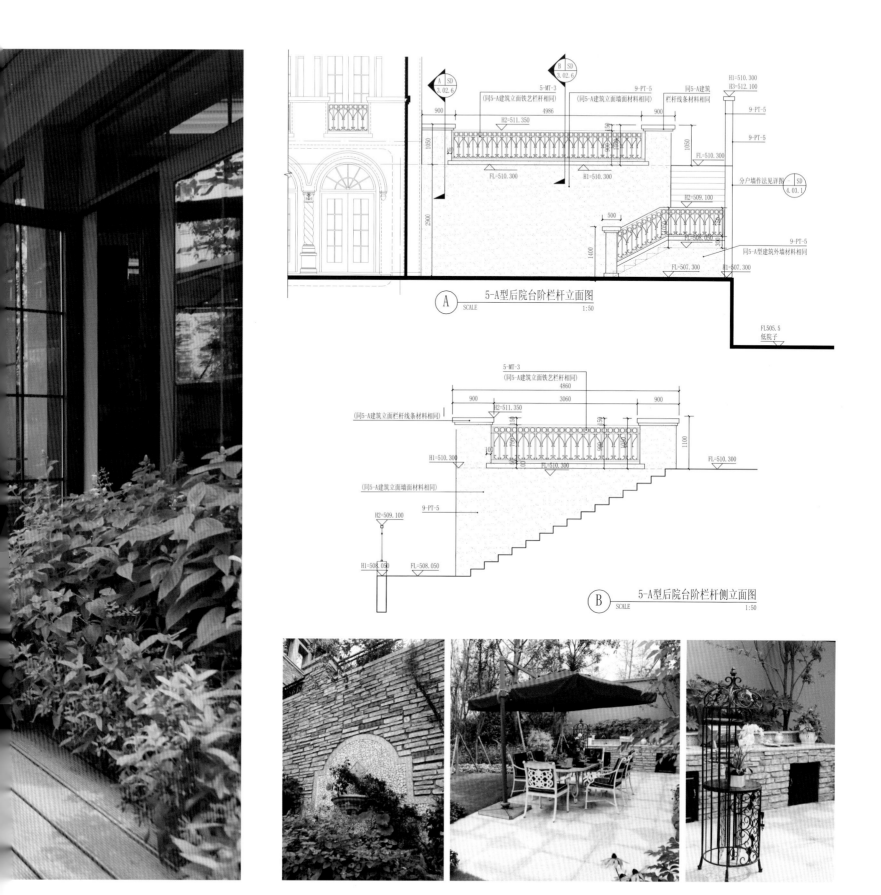

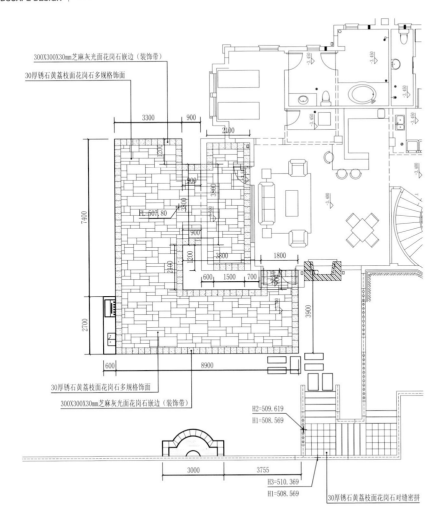

■ 简洁的汀步直通向屋内
Brief step stones on water lead directly to the inside of the room.

■ 浪漫的法式风格休闲空间 Romantic leisure space in French style

■ 一草一木，一桌一椅营造出如此优雅的庭院休闲空间
  The grass, trees, tables and chairs create such an elegant courtyard leisure space

■ 鲜艳的花朵点缀着茵茵绿草，在这里喝杯下午茶与朋友谈心是那么惬意   Colorful flowers decorate the green grass. It is such a delight to talk with friends over afternoon tea

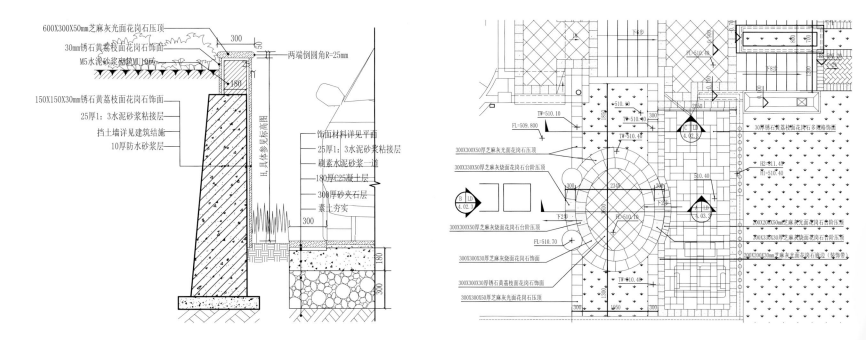

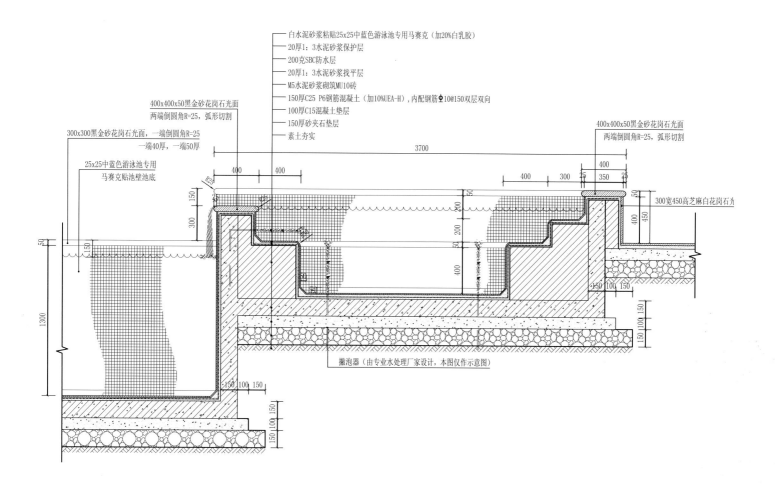

- 白水泥砂浆粘贴25x25中蓝色游泳池专用马赛克（加20%白乳胶）
- 20厚1:3水泥砂浆保护层
- 200克SBC防水层
- 20厚1:3水泥砂浆找平层
- M5水泥砂浆砌筑MU10砖
- 150厚C25 P6钢筋混凝土（加10%UEA-H），内配钢筋Φ10@150双层双向
- 100厚C15混凝土垫层
- 150厚砂夹石垫层
- 素土夯实

400x400x50黑金砂花岗石光面
两端倒圆角R=25，弧形切割

300x300黑金砂花岗石光面，一端倒圆角R=25
一端40厚，一端50厚

25x25中蓝色游泳池专用
马赛克贴池壁池底

400x400x50黑金砂花岗石光面
两端倒圆角R=25，弧形切割

300宽450高芝麻白花岗石光

撇泡器（由专业水处理厂家设计，本图仅作示意图）

别墅景观设计 | 龙湖·长桥郡

VILLA LANDSCAPE DESIGN

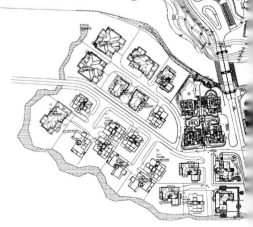

婉转山河，惊艳绽放

龙湖·长桥郡

Devious Rivers and Mountains, Amazing Blossom

Longhu, Long Bridge County

开发商：	成都龙湖锦城置业有限公司
项目地点：	成都新津花源镇
总占地面积：	约800亩
景观面积：	约30万平方米
项目类型：	纯独栋别墅
设计单位：	成都绿茵景园　设计一所
主创团队：	潘旭、潘迪、粟凡粒、姚抒雅、阙龙庆、肖浩波、雷冬
设计及施工时间：	2008年至今

设计主题 Design Theme

龙湖·长桥郡位于距离成都仅28公里的牧马山片区，项目位于新津花源镇，距离航空港仅18公里。总占地面积约800亩，规划为纯独栋别墅，融汇了岛屿、溪岸、山地三种各具特色的别墅风格。环境优美，气候宜人。项目地块毗邻四川国际高尔夫球场。杨柳河、碾河在此交汇后沿着牧马山麓顺势而下，别墅沿着缓坡蔓延，与林荫、道路高低错落，共同形成了纯粹的自然风景。

小区主干道、会所区、样板区以及一期现已完成建设，一致受到外界的好评。龙湖·长桥郡被称为"原著山水别墅"，以一山（背倚牧马山）、两河（杨柳河、碾河环绕）为优秀自然条件基础，打造稀缺纯独栋别墅区。按照规划道路等级将小区的车行道路分成三个级别。主干道为东西向贯穿的一期用地，从入口处直通会所区域的一条林荫大道。现已完成建设，营造出一条古树参天，幽静深远的回家之路。

项目风格的定位是根据该项目本身建筑风格的定位及场地分析结论而确定。英式庄园风格骨子里透露出的一种没有任何杂念的朴实、简洁大方、淡雅清新，让久居都市、终日劳碌的人们更能感受到一种宁静和舒适，回归自然的情趣。

本项景观设计风格主要通过林荫道、水体、桥、植物、入口标识四种元素得以体现。

The Longhu Long Bridge County is located in the Muma mountainous area, only 28 kilometers away from the Chengdu. The project is located in New Jinhuayuan Town, only 18 kilometers away from the airport. The project covers a total area of about 800 mu. It is planned for separated villas in combination with the three distinctive villa styles of islands, shores and mountains. It has a beautiful environment and a pleasant climate. The project site is next to Sichuan International Golf Course. The Willow River and Nian River flew along the foothills of Muma Mountain. The villas that spread along the gentle slopes interlocking with the boulevard and roads present pure natural scenery.

The construction of the main road of the community, the club area and the sample area has been completed with praise from the outside world. The Longhu Long Bridge County is called "original villas with mountains and rivers". It creates a rare single-family villa area based on excellent natural condition that it leans against one mountain (Muma Mountain) and is surrounded by two rivers (the Willow River and Nian River). According to the road level planning, the community is divided into three levels. The main road is a boulevard from the north to the south runing through the first-stage area from the entrance to the club area. Now the main road is completed and it creates a peaceful and spreading road leading home.

Project style is chosen according to its architecture style and conclusion of its site analysis. British style reveals its simplicity without distracting thoughts, conciseness and freshness. It brings urban people busy working all day long an affection of seclusion, comfort and returning to nature.

The landscape design style is presented by four elements of the boulevard, water, the bridge, plants and the entrance sign.

■ 现场特意保留的原生大树 Primary Trees Specially Reserved in the Site

■ 在干净的草坡上欣赏对岸的风景，是对杨柳河绝佳的诠释
Appreciate the landscape across the bank on the clean grassland and you will know what Willow River looks like

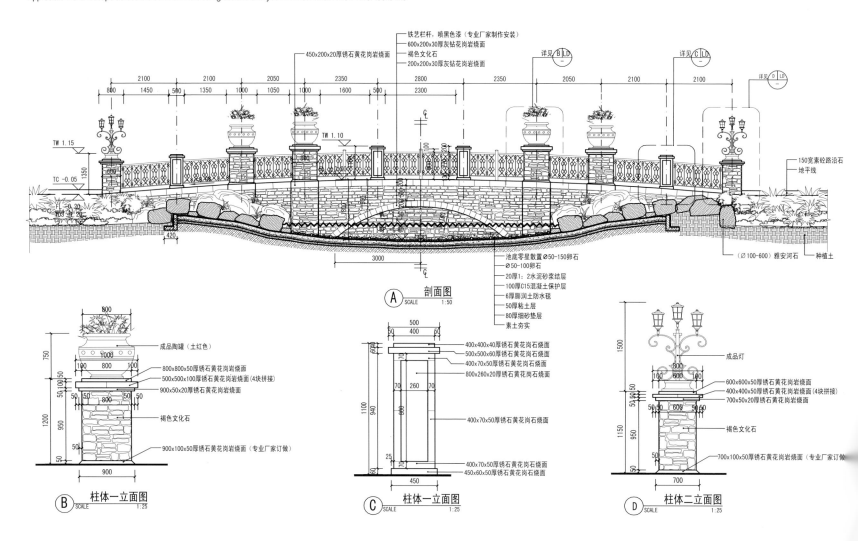

① 林荫大道 Boulevard

运用植物造景，营造出具有浓厚田园风格的道路景观系统，主要道路以多种冠大荫浓的大乔木为主，宅间道路则选择花色淡雅的开花乔木为基调，辅以色页树种及花灌木，下层为小灌木及地被，常绿与落叶相结合，层次丰富，颜色秀丽，营造幽静深远的隐私感，具英式乡村生活特色。

It uses plants to create a road landscape system with a strong flavor of rural life. On the main road, there are mainly lush and tall trees. While on the roads in the residence, the roads are decorated with trees with simple and elegant flowers as a basis and the lower level is decorated with small shrubs and earth level. The combination of green trees with fallen leaves reveals diversified levels and beautiful color. It also creates a quite and lasting atmosphere of privacy with a British rural life style.

■ 无边界泳池 Boundless Swimming Pools

② 水体 Water

水是本项目景观设计中较为重要的元素，宅间水溪既是别墅区绿化的核心景观，也是庭院分隔的一大要素。瀑布、溪流、跌水、水中小岛与植物等其它景观元素相配合，构成独特的景观空间。利用现状河流，在满足防洪的基本要求上，以草坡和乔木为主，形成视线深远、开阔的景观空间。

Water is an important element in landscape design of this project. The streams among the houses are not only the core landscape of the villa area, but also an important element for separating the courtyards. The falls, streams, water plunge, the island in the water and plants in combination with other elements of the landscape constitute a special landscape space. On the basis of meeting the requirement of flood control, the spreading and open landscape space is formed mainly with grass slopes and trees.

③ 桥 Bridge

集淳朴自然于一身，富有英式田园特色的石桥，构成别墅区特有的一道风景线，强化"长桥"主题。

The bridge combines simplicity, nature and British rural life style and it is a special scenery line in the villa area. It also strengthens the theme of "Long Bridge".

■ 会所前的花儿亦如天一样的蓝
The flowers in front of the club are as blue as the sky

④ 植物 Plants

植物是景观设计中重要的景观元素，它包括草坪、乔灌组合的植物群落等，强调舒适宜人的休闲体验，在项目中慎用经过整形修剪的植物。植物景观应与其他的景观元素，如水体、挡墙、地形等相配合使用。

The element of plants is very important for landscape design, including grassland and group of bushes. Plants design stresses pleasant leisure experiencing and the trimming plants are carefully used. The plants should match other elements, such as water, retaining wall, terrain, etc.

⑤ 入口标识 Entrance Sign

每个组团、每个户型都有属于自己的标识物，形成典型的景观符号，给人以强烈的归属感。

All the groups and house types have their own signs to form their typical land-scape symbols which give people a strong sense of belonging.

■ 山河之间，尽归独户前院
In the front yard of the house, you can enjoy the scenery of mountains and rivers.

林中漫步
龙湖·江与城·原山

Walking in the Forest
Longhu, River and City, the Original Mountain

名　　称：江与城三期示范区
所在地：重庆渝北区
投资额度：600000万元
工　　期：2008.10—2009.5
设计公司：重庆蓝调国际（绿茵景园集团公司）设计二所
设计总监：张勇
项目负责：王轶
设计团队：陈云川、艾珍、潘启渝、刘相梅
规　　模：12000平方米
主要材料：枕木、花岗岩、陶土砖、碎石
主要植物选择：白桦、桂花、红叶李、藤七菊、杜鹃

设计主题 Design Theme

重庆的丘陵地貌构成了其独特的山地景观优势。许多设计项目利用场地竖向因地制宜，提出丰富的自然山地空间概念。其中渝北区江与城居住片区，依山傍水，使其成为最受欢迎的居住环境和城市开放空间。

江与城第三期别墅区样板段用地与北面市政道路存在10米高差，两者间的市政绿地成为进入样板房的步行入口。设计师受重庆龙湖集团委托，希望很好地利用该绿地来解决竖向问题，使其成为具有吸引力及自然舒适的步行环境。设计初期，设计师在分析了其功能和流线之后同开发商一起构思了几种入口的氛围：有展现别墅区入口华贵、气派的轴线对称式布局；有随场地等高线走势，运用优美、流动的自由曲线来疏理林下的步行空间；还有一种是想尝试更接近自然的处理手法，远远望去，展现在第一视线的景色是干净、舒缓的草坡，建筑在草坡的背景树林后含蓄地显露出来。步道从路边人行道转角处进入草坪，在草坡一侧的林间弯延曲折地向高处伸展。

第三个构思成为最终深化的方案，但自然的景观营造却成为最难创造个性的选择。因此设计师在构建起入口空间关系和环境氛围之后，把重点放在了植物的营造和步道材质的选择上。自然的石材按照一定规格形式来拼贴，再根据场地高差关系，把控好台阶与地形的节奏感，这样并不难营造出既现代又自然的步道肌理效果。但这样的处理方式还达不到令人满意和惊喜的程度。因此设计师又通过自己的生活经验，运用回收的材料"枕木"，作为主要的饰面，让这种经过特殊处理后的木质回归林间，同时用白色的碎石垫层来衬托质地厚重、色泽自然的枕木，让木材有序无意地铺垫出一条通向自然的林中步道。

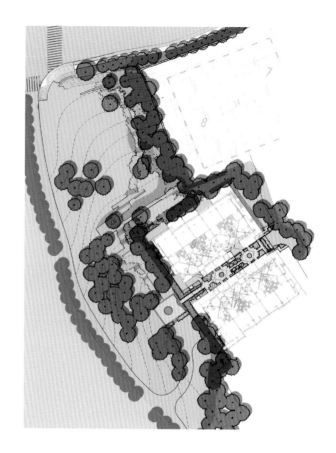

■ 高差的解决成为此次设计的重点难点。步道沿草坪而设，蜿蜒曲折的沿草坡而上
Solving the problem of height difference is the focal and difficult point of this design. The footpath is designed along the grassland and winds up along the grass slope

别墅景观设计 | 龙湖·江与城·原山
VILLA LANDSCAPE DESIGN

The hills in Chongqing constitute its unique advantages of hilly landscape. Many projects are designed according to the specific location and the concept of rich natural mountainous space is proposed. River and City Community in Yubei Region is constructed close to the mountain and the lakes, making it the most popular residential place and urban open space.

The villa region of the third-stage River and City Community has a 10-meter height difference in the sample section land and the northern municipal road. The municipal green area between them becomes the entrance of the sample house. The designer entrusted by Chongqing Longhu Group hopes that the green area can be used to solve the vertical problem and can be made into a fascinating and pleasant walking environment. In the early design, the designer with the developers worked out several kinds of atmosphere for the entrance after analyzing the green area's function and streamline. They considered the axis symmetric layout presenting the entrance's magnificence and luxury or using an elegant and flowing free curves trending with the height of the site to arrange the space under the tree. They also came up with a dealing way to try getting closer to nature. Seen in distance, the neat and gentle grass slope comes into fight first. The background forest constructed in the grass slope is revealed in a reserved way. The road enters the grassland from the corner of the passenger way and spreads to the high place in a winding way.

别墅景观设计 | 龙湖·江与城 ·原山
VILLA LANDSCAPE DESIGN

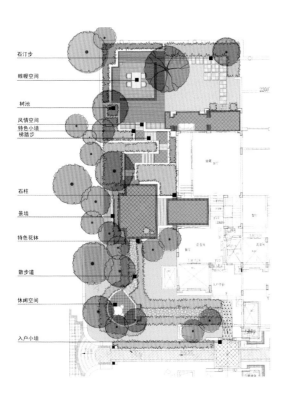

石汀步
帷幔空间
树池
风情空间
特色小墙
梯踏步

石柱
景墙

特色花钵

散步道

休闲空间

入户小墙

■ 精致的小空间和细节处理使得整个场地更加有品位，贴心的枕木步道、陶瓷花盆、休闲坐椅，让整个场地高贵而又不与人疏离
Exquisite little space and detail dealing promote the grade of the whole site. Considerate sleeper footpath, pottery flower pots and leisure chairs make the whole site noble but not alienating people

The third idea becomes the final deepened plan, but the creating of natural landscape becomes a choice which is hard to create characteristics. Therefore, the designer puts emphasis on the creation of the plants and the selection of roads material after constructing the entrance space relation and the environmental atmosphere. Natural stone material is put together according to a certain standard. Then the rhythm between the stairs and the site should be well controlled according to the height relation of the sites. Thus, it is not difficult to create a modern as well as natural effect. However, this approach is not yet the pleasant or amazing one. Therefore, the designer works out a way based on his life experiences to use recycled material "sleeper" as a main decorating surface. The wood after such processing returns to nature. At the same time, the filling level of the white broken stones serves as a folio to present the high quality and natural color of the wood. In this way, the wood paves a road leading to the natural forest.

在林中植物的设计上，设计师提出了大胆的想法，引用北方的白桦作为主干树种。对场所而言，其个性被鲜明地展现出来，而在植物移栽的课题上，北方植物到南方区域性的移植也做了一次重要的尝试。

In the design of the plants, the designer proposes the adventure of idea that the birch trees should be served as the main trees. In this way, the site's character is presented in a bright way. For plants transplanting, transplanting plants from the north to the south is an important attempt.

别墅景观设计
VILLA LANDSCAPE DESIGN
海航·香颂湖国际社区

■ 上千亩大型国际社区整体鸟瞰
Overall Aerial View of Large International Community of Thousands of Acres

→

汤泉湿地，原野景观
海航·香颂湖国际社区

Tangquan Wetlands, Wilde Landscape
Sailing, chanson lake international community

开发商：四川海航鸿景实业有限公司
项目地址：四川都江堰翠月湖公园侧
规划用地面积：97.38公顷
前期设计面积：意境区约10万平方；启动区约6.57万平方；
项目类型：大型社区住宅楼盘（公建、洋房、独栋、联排）
景观设计单位：成都绿茵景园　设计三所
项目负责人：肖浩波
主创设计：曾旭
设计团队：杜佩娜、白锐、潘宇、雷冬、潘强、彭溟鸥
设计及施工时间：2009至今

■ 林荫大道——光影婆娑　Boulevard-Whirling Light And Shade

■ 气势磅礴的社区主入口手绘效果图
Manual Effect Picture of Main Entrance of the Community of Great Momentum

设计理念 Design philosophy

"乐活——LOHAS"即健康、可持续发展的生活方式。

→ 设计阐述 Design Explanation

拥有上千年历史的都江堰，其源于山脉且分支众多的河流和运河体系是该地区景观开发的模范榜样。珍惜爱护资源是我们最重要的开发原则。我们的目标在于实现一个持久的、与当地自然环境相和谐的高质量生活。

设计从MUDI的规划开始，在经过现场的勘测，实地的考察之后，充分整合现有资源，场地特性，最大限度的保持现状，利用现有优势：穿过该项目的三条水系、充分的水资源、天然湖泊——香颂湖，使项目做到纯粹、天然、生态、原野。

设计以"水"为主线，贯穿整个项目，从前期意境展示区到后期独栋、联排都以水为契机，贯穿始末。整个社区场地北高南低，南北贯穿五条水系，以中央公园水系最为宽阔，河道尺度控制在20-50米，而环绕岛屿状独栋别墅的分支水系，更是将建筑进行包围，成为大型水上别墅群。水系从北至南流经社区的各个组团，最后到达核心区域——香颂湖，宽阔的湖面，摇曳的芦苇，自由飞翔的水鸟，千亩花海的神话在这里展开……

Dujiangyan has a history of a thousand years. Its innumerate rivers and canal systems which originate from mountains and have many branches are the examples of landscape development in this area. Caring and protecting resources is our most important development principle. We aim at achieving a lasting and high quality life in harmony with the local natural environment.

The design starts from the plan of MUDI. It integrates the existing resources and site features to the greatest extent after surveying the scene. The design maintains the current circumstance to the greatest degree and uses the existing advantages which are three water systems through the project, adequate water resources and natural lakes (Chanson Lake). The project achieves the features of purity, nature, ecology and wilderness.

香颂湖居
像一推门，
青城远山成了楼台中的画。
长廊履步，在香气波光里，
踏断夏日的箫声，
恍如山林间被樵夫惊散的鸟鸣。
湖畔，大朵大朵的荷花。

独坐这水榭
为着身后石笋上苍玉泪痕……
……昨夜，旱舫里小红用琵琶
模仿浔阳江头的秋水……悠悠

思量明年，该问友人
要若许花种：香颂湖畔熏衣紫
普罗旺斯，在梦中惊艳……

■ 次入口效果图
Manual Effect Picture of Secondary Entrance

景观设计根据建筑的风格及业态进行专项设计，在整体"原野"风格的统筹下，严格按照规划设计的原则进行景观设计：

①"香颂湖景"——意境区
Chanson Lake Scenery - Artistic Area

香颂湖以其开阔的水域和对公众开放的河岸成为社区内的一大亮点。现代的会所建筑，开朗大气的意境区景观场景，场景式休闲设施，浪漫的原创景观风格。

The landscape is specially designed according to the architecture style and condition. It is designed in strict accordance with the principles of landscape design under the overall planning of a "wild" style.

Chanson Lake has a hilight of open rivers and public bank. There is modern architecture, wide scenery in the artistic area, leisure facilities in the scene and romantic original landscape style.

The design takes "water" as the main line throughout the entire project. The pre-stage conception display area, post-stage detached houses and town houses all take water as an opportunity to run from the beginning to the end. The whole community site is higher in the north and lower in the south. There are five river systems running throughout the community, among which Central Park system is the widest with a wide control of 20-50 meters. What's more, the branch water systems around the island-like detached villas surround the buildings. Thus large-scale groups of villas on water are formed. The rivers flow through groups of the community from north to south and finally reach the core region -Chanson Lake. It is a broad lake with swaying reeds and free flying birds. There is the sea of lowers flourishing besides the lake.

■ 极具原野风情的会所后场阳光草坪　Sunshine Grassland with a Strong Wild Flavor Behind the Club

■ 占据最佳视野的会所前场展示区手绘效果图　　Manual effect picture of exhibition area in front of the club enjoying the best view

■ 曲折的花径，将游人的脚步及视线引向对岸高低起伏的高尔夫草坪
Winding road with flowers directs tourists' footsteps and sight to the undulating golf grassland

■ 大型水上公园，林溪曲径，原野质朴
Large parks on water, forest streams and winding roads are in a wild and plain style

②"水上公园"——中央公园区
Park on the Water - Central Park District

整个中央公园沿宽20-50米的河道进行展开，宽阔的河道，为两岸社区提供宽阔的视野。简练的植物配置和大尺度的水面展现了大气的景观设计，贯穿南北的水系演绎着纯粹的湿地景观。

The overall Central Park is developed along the river with a width of 20-50 meters. The wide River provides a broad vision for the community along the bank. Concise match of plants and largescale of water present the landscape design with great momentum. The rivers running from the north to the south interpret the pure wetland landscape.

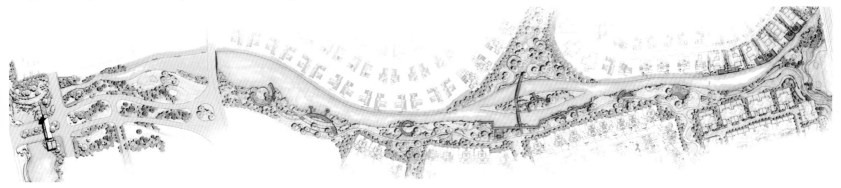

■ 中央公园平面图 Plan of Central Park

■ 质朴的木桥，清澈的河水，开敞的草坪，一切这么宁静自然，让人不由得归心似箭　The bridge in a plain style, clean water and open grassland are so quiet and natural that people are anxious to return home.

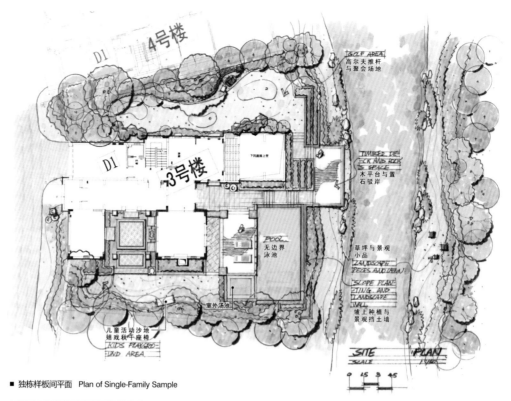

■ 独栋样板间平面　Plan of Single-Family Sample

■ 独栋别墅被绿化环绕，以植物隔离取代围墙，让人和自然更亲近
Single-family villas are surrounded by green landscape. Plants isolation in spite of walls makes people close to nature

③ "湖中小岛"——独栋别墅区
"Island in the Lake" - Detached Villa Area

　　被河流以及绿化环绕的别墅小岛以平缓的建筑体块高度从地平线升起。别墅小岛中的独立别墅创造了最高规格的私密性、舒适度和安全性。每个别墅小岛都有属于自己的形态主题。

　　The villas island surrounded by the rivers and the greening rises from the horizon in a gentle height. Independent villas on the island have created the highest standard of privacy, comfort and safety. Each villa has its own form theme.

■ 以异域特色的越南佛像结合阔叶和彩色植物，颜色绚丽的室外和富有生活情趣的家具，来体现越南的法式风情。

■ 跨桥而过，临湖而居　Crossing the Bridge and Living Besides the Lake

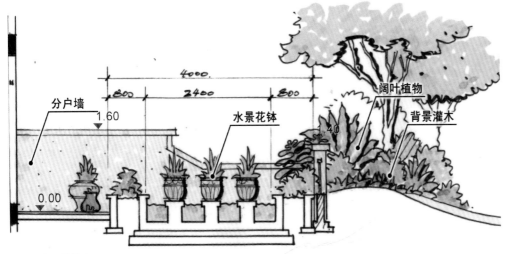

■ 庭院小水景手绘剖面图
Manual Profile Drawing of Little Water Landscape

④ "鲜花盛开的林荫大道边的住宅区"
——联排别墅区

"Residential area besides the boulevard with flourishing flowers"
——Townhouse Area

中央住宅区得益于它优越的地理位置，介于中央公园以及长满树木和开满鲜花的林荫大道之间。林荫大道整体景观的打造，通过道路曲线及植物的配置，从林冠线及林缘线几个方面进行景观阐述，表达出疏朗大气的景观设计意图，增强引导性，延伸景观层次。

The central residential area benefits from its predominant geological location. It is located between the central park and the boulevard with flourishing trees and flowers. The overall landscape of the boulevard is created by the curve of the road and the arrangement of plants. The landscape is developed from the canopy line and forest edge line. It expresses the open and splendid landscape design idea, strengthens the role of guidance and spreads the landscape levels.

■ 大气简洁而又充满异域风情的庭院空间 Splendid and simple courtyard space with an exotic flavor

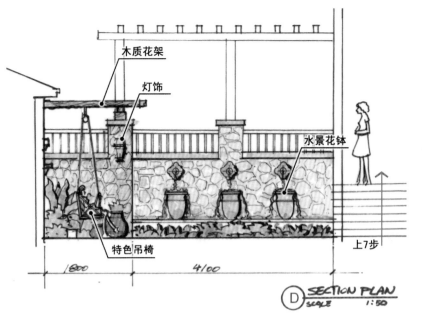

■ 独栋别墅大气的私家庭院，丰富的空间打造，满足业主所有梦想的私享领域
  There is a splendid private courtyard and abundant space in the single-family villa which satisfies the owner's desire for private space

■ 庭院休闲空间，享受阳光下午茶 In the leisurely space of the courtyard, you can enjoy the sunshine afternoon tea

复地·别院

Fudi-House by the Park

01 水景广场
02 童话森林
03 金色大道
04 原野芳歌
05 华尔兹广场
06 花溪烂漫
07 幼儿园
08 法桐大道
09 玫瑰花园
10 葡萄园
11 水影叠花
12 紫薇牵藤
13 中心绿地
14 棕榈泉
15 组团绿地
16 蓝湖圣景
17 千层花境
18 樱花园
19 组团绿地
20 古藤花影

项目开发商：重庆润江置业公司（上海复地集团）
项目地点：重庆经开区金开大道
项目规划用地面积：5.5万平方米
景观面积：3.8万平方米
项目类型：新古典主义风格居住区
设计单位：重庆蓝调国际（绿茵景园集团公司） 设计二所
主　创：王轶
团　队：敖翔、卢燕、刘盛义、艾珍、潘启渝
设计时间：2008年

项目简介 Project Theme

2007年重庆蓝调城市景观规划设计公司为重庆市金开区复地上城二期居住环境提供了崭新的设计方案，设计师确立了以一种浪漫、休闲、自然，倡导维也纳新古典主义的风格作为基调，融入人们的生活。而维也纳就是这样一座以音乐盛名，并弥漫着皇家气息让人魂牵梦萦的城市。我们的工作也就是在这醉意的氛围中，谱写着"复地别院"这篇灵动的乐章。

水景、植栽、小品与场地北面的高尔夫球场是复地二期景观的亮点。

设计师利用地形高差变化形成的谷地，营造出一条串接住区的沟谷溪流花园，构成整体的住区公共休闲骨架。缓坡与植物共同营造出自然的坡岸，步道与水面相互交融、时隐时现，重点处的临水场地在高大乔木植物掩映下，使得活动空间轻松而宜人。住宅间的环境更加突显设计艺术的细腻与精致。在主轴的大草坪上，一列音乐的天使雕塑奏响着舒缓的乐章，此处整洁的植栽同溪流和它间的绿化形成疏密的搭配，密处"密不透风"，疏处"疏可跑马"，既有密林清幽的郁郁葱葱，又有草坪开阔的坦坦荡荡。这里，水是清澈的，天空是蔚蓝的，气候是宜人的，还有洁净的草坪，蜿蜒的水岸在斑驳的树荫的点缀下依偎着自然的栈道。傍晚，温暖的泳池映射着绚烂的阳光，散布在岸边伞下美酒的醇香飘散出醉人的悠闲与皇家浪漫的情怀。

无论是自然风景，还是人工小品，或是两者间的搭配与衬托，无不雅质、唯美。独特的异国风情，浓浓地表达在每个细节之上，从建筑的形式到窗户的花纹，从灯饰的造型到标识体系的风格，都构建出整体环境定位的统一、和谐与不凡。

■ 千层花境透视
Perspective of Flower Landscape with Thousands of Layers

Project Theme

In 2007, Chongqing Blue Tone Urban Landscape Planning and Design Company provided a new design plan for the second-stage Fudi Upper City in Jinkai District in Chongqing City. The designer used a romantic, leisurely, natural and Vienna new classical style as a fundamental key which combined with people's life. As Vienna is famous for music, it is filled with royal breath that fascinates people. Our work is to compose an appealing work of "House by the Park" in such an intoxicating atmosphere.

Waterscape, plants, sketches and the golf course in the north are the highlights of the second-stage Fudi landscape.

The designer uses the valley formed by terrain elevation changes to build up a gully stream garden which leads to the living region. The garden helps the community form a public frame of entertainment in living place. The space is made relaxing and delightful by the banks formed by the combination of the gentle slopes with plants, the harmony of roads and water and its occasional appearance and the waterfront site decorated with tall trees. The environment of the residence region gives prominence to the delicacy and exquisiteness of design art.

■ 金色大道透视图 Perspective Drawing of Golden Avenue

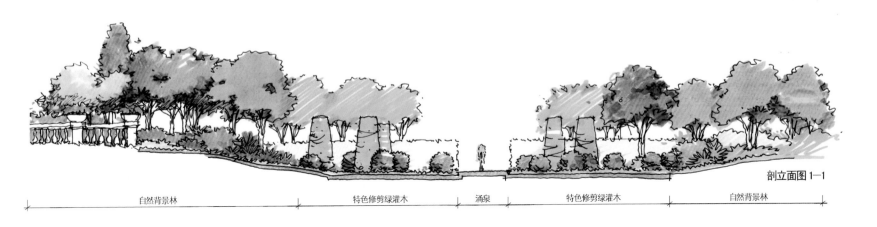

剖立面图 1—1

| 自然背景林 | 特色修剪绿灌木 | 涌泉 | 特色修剪绿灌木 | 自然背景林 |

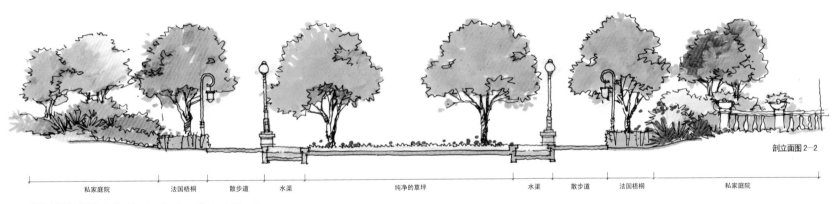

剖立面图 2—2

| 私家庭院 | 法国梧桐 | 散步道 | 水渠 | 纯净的草坪 | 水渠 | 散步道 | 法国梧桐 | 私家庭院 |

■ 童话森林剖立面图 Profile Elevation Drawing of Enchanted Forest

On the main large grassland, there is a list of the angel statues with soothing music. There, neat plants, streams and the greening among the houses form crowding or spacious places. In the crowding places, wind can hardly come through, while, in the spacious places, horses can run through. The dense forest as well as the open grassland can be seen there. There is clear water, blue sky, pleasant climate and neat grassland. Winding waterfront is lying next to the natural plank road decorated with uneven shade. At dawn, the sunshine shines upon the swimming pool. And the nice smell of the wine under the umbrellas scattered along the bank fill people with fascinating pleasure and royal romantic feelings.

The natural scenery, artificial decoration and the combination of them are all elegant and beautiful. The unique exotic amorous feelings are expressed in all details. The styles of the architecture and symbol system, the patterns of the windows and lights constitute the integrating, harmonious and uncommon environment.

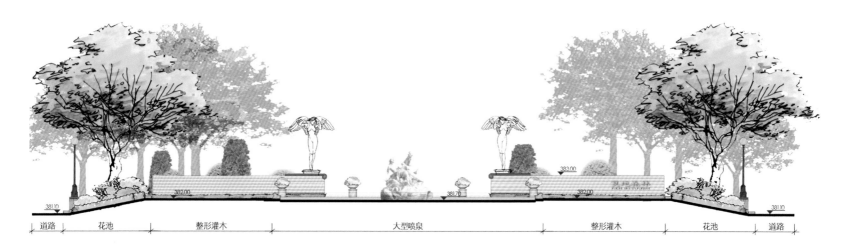

| 道路 | 花池 | 整形灌木 | 大型喷泉 | 整形灌木 | 花池 | 道路 |

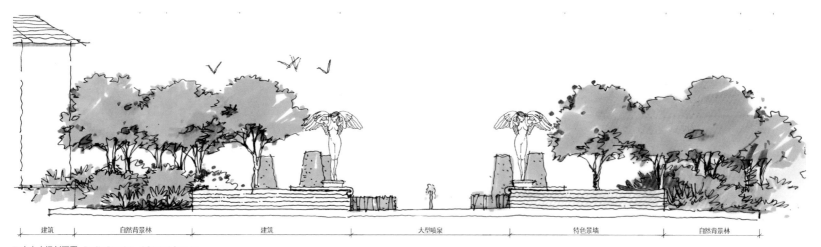

| 建筑 | 自然背景林 | 建筑 | 大型喷泉 | 特色景墙 | 自然背景林 |

■ 中心广场剖面图　Profile Drawing of Central Square

■ 节点广场剖面1　Profile Drawing 1 of Joint Square

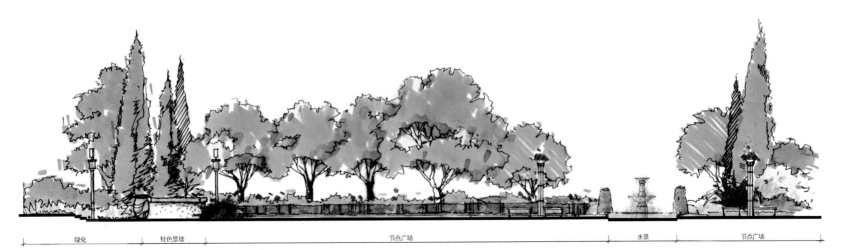

| 绿化 | 特色景墙 | 节点广场 | 水景 | 节点广场 |

■ 节点广场剖面2　Profile Drawing 2 of Joint Square

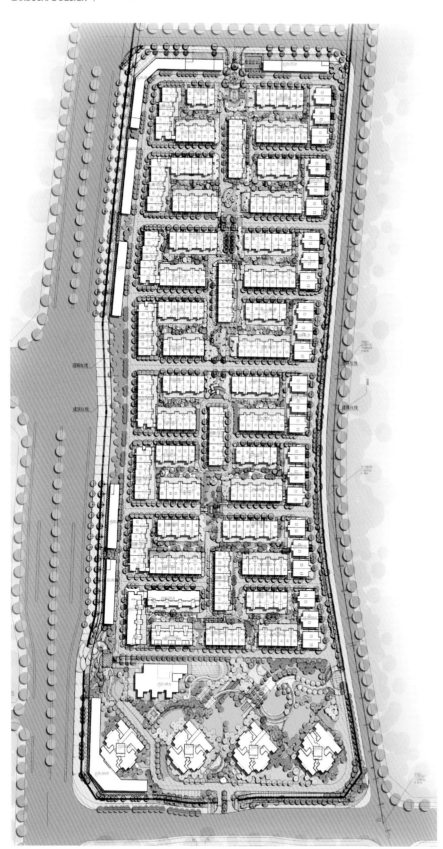

梦幻之家
富力城·维多利亚庄园
Dream of Home
Fuli city, Victoria Manor

项目名称：重庆富力城C区（维多利亚庄园）
项目类型：英伦风格居住区
项目面积：规划用地11.2万平方米
景观面积：4万平方米
设计单位：重庆蓝调国际（绿茵景园集团公司）设计三所
项目地点：重庆大学城
委托单位：重庆富力城房地产开发有限公司（富力地产集团）
主　创：肖勇、李锦香
团　队：熊扬眉、杨懿、左春丽、张骊瑜、郭怿、彭盼
设计时间：2009年

项目简介 Project Description

位于重庆大学城的富力城C区为英伦风情的联排别墅，项目密度较高，除去私家花园及道路，留给景观的空间场地非常有限。设计师根据建筑的规划布局，以英伦风格的景观小品，铺装形式和色调，结合田园的植物空间，将整个空间归纳为中央景观轴线（景观通廊）及四大景观节点。以"维多利亚庄园"的景观主题为设计理念。

入口景观、特色花园、自然田园的植物植栽成为维多利亚庄园的景观亮点。

There are town houses with British flavor in C Department of Fuli City on Chongqing Campus. The project has a higher density and there is very limited space for landscape except for the private gardens and roads. According to the construction design of the architecture, the designer lays out the form and color with landscape sketch in British style. The designer also combines the space of plants to centralize the overall space as a landscape axis (landscape corridor) and four landscape points. It is designed with the concept of Victoria Manor.

Entrance landscape, the feature garden and plants of the natural fields are the landscape highlights of Victoria Manor.

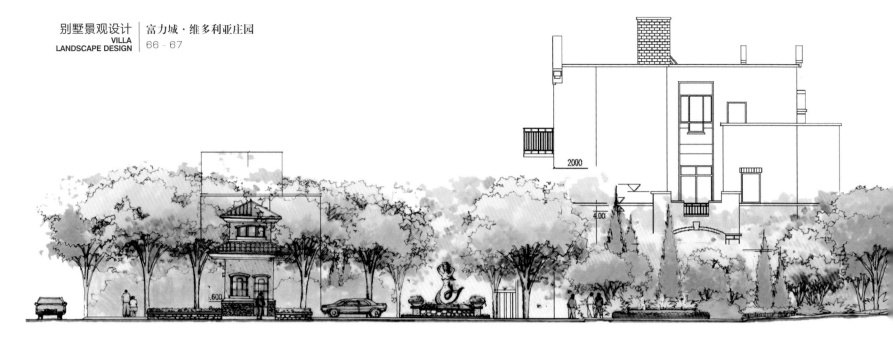

① 入口景观区 Entrance Landscape Area

人行入口，设计师因地制宜，利用地形高差变化及周边场地的环境关系，营造流线型台地式的入口。场地与地形高差紧密锲合，以自然石材的台地花池，花砖立面的台阶，温婉地展现出英伦田园风格的优美画面。

车行主入口，英伦风情的门廊塔楼建筑形态，倒影在镜面水景中，在竖向上增添了场地恢宏的气势，突显庄园式花园的入口尊贵品质风貌。入口注重人车分流的形式和车行环岛的设置，体现人性化设计特点。沿着岗亭与雕塑小品构成的对景轴线，将人们的视线引导至庄园的深处，使景观层次幽深而富有的变化。

The Pedestrian Entrance: The designer adjusts measures to local conditions and takes advantage of the connection between terrain elevation changes and the surrounding environment. The entrance is designed in a streamlined form. The close combination between the environments with the terrain elevation, flower pond made of natural stones and the steps of the tiles mildly show a beautiful picture in British rural style.

The Main Entrance for Cars: The porch tower in British style is mirrored on the water, which adds momentum to the site and presents the noble qualities of garden entrance. The design of the entrance pays attention to the separation between passengers and vehicles and the setting of encircled island for cars. Along the axis constituted by the sentry box and the sculpture, people's sight is led to the landscape of the manor, diversifying the landscape in various layers.

② 特色花园景观区 Feature Gardens Area

庄园内，以主景树木为主题，分别设计为梧桐、银杏、桂花、香樟四个特色花园。

梧桐花园，以12M以上精选法国梧桐为主打特色植物，在进入庄园以环抱形式种植，呈现绿意葱茏、树影婆娑的美妙图景，结合背景梧桐林下的休息平台和背景墙，构成庄园内的视觉焦点。

银杏花园，带状的精致水景，挺直的银杏，布置于园路两侧，造型景墙的潺潺流水，显出花园的静谧氛围，为住户带来惬意安然的生活空间。

桂花花园，根据地形的高差变化，设计起伏变化的林中草地。点缀各色桂花，成为特色的芳香花园。

香樟花园，以香樟为主题植物，围合节点场地，形成环境宜人的休息空间。

各个花园各具特色，步移景易，空间开合有致，在蜿蜒曲折的穿插路径中，感受景观带给人们的变化之美。

充足的绿化，是宜居环境的基础，再赋予精致动人的景观风格细节，将带给人们富有的文化内涵和气质。富力城C区"维多利亚庄园"，正是本着宜居与人文的特性，而精心设计的住区景观。

The manor is designed with the theme of scene trees and is divided into four feature gardens respectively with the themes of phoenix trees, ginkgo trees, osmanthus trees and camphor trees.

Phoenix Garden is featured with French phoenix trees which are 12 meters high. They are planted in the way of surrounding in the manor and take on a beautiful picture of bushy greenness and whirling shadows. Combined with rest platform and the wall in the background of phoenix trees, the garden constitutes the visual focus in the manor.

Ginkgo Garden has a delicate ribbon-like water scenery. The erect ginkgo trees are planted on both sides of the garden roads. Besides, there are murmuring streams under the landscape wall. All these reveal a quiet atmosphere and bring pleasant living space for residents.

Osmanthus Garden is designed with undulate grasslands in the forest according to the terrain elevation changes. It is decorated with various osmanthus flowers which make it a fragrant garden.

Camphor Garden is featured with camphor trees. The trees surrounding the joint sites form a pleasant rest space.

■ 疏朗大气的主入口形象　The open and splendid image of main entrance

别样的白色之城

湖南长沙橘郡·米哈斯小镇

Special White City
Orange County in Changsha, Hunan, Mijas Town

所在地： 湖南省长沙市芙蓉南路
投资额度： 1100万元
工 期： 2008.11 – 2009.5
委托单位： 湖南橘郡城镇置业有限公司
设计单位： 重庆蓝调国际（绿茵景园集团公司）设计三所
施工单位： 重庆绿茵景园园林工程有限公司
项目总监： 张坪
设计团队： 肖勇、敖翔、郭怿
景观面积： 22000平方米
主要材料： 欧洲砖、花岗岩、陶土砖、花砖、陶瓦、马赛克瓷片、手扫漆。
主要植物选择： 榉树、桂花、香樟、栾树、棕榈、红叶李、碧桃、山茶、杜鹃、肾蕨

设计主题 Design Theme

橘郡·米哈斯项目的地理环境和建筑规划设计，以西班牙国宝级小镇Mijas（米哈斯）为蓝本，试图带给人们别具情调的异域生活感受。原生态的丘陵山地、静谧的湖泊、葱郁的植物生命，精致热情的西班牙风情建筑和统一的景观细节，构成了橘郡迷人风景的核心基础。从景观设计到整体施工建设，从环境氛围的营造到细节造型材料的选定，都始终紧密围绕西班牙风情特色而展开。

The geographical enviroment and the architecture of Orange County—Mijas Town project are designed based on the famous Mijas Town in Spain, aimed at bringing people exotic life feelings. There are natural hills, quiet lakes, lush plants, delicate and passionate architecture in Spanish style and integrating landscape details, constituting the core basis of Orange Town's fascinating scenery. The landscape design and construction, the environmental atmosphere creating and the detail material selection are all centered on the Spanish style.

① 欣赏的景观 —— 依山而就的标志性入口
Appreciating Landscape
— Symbolic Entrance Built along the Mountain

项目选址位于长沙市郊的城市快速干道一侧，入口如何形成打动人心的标志景观，成为设计的重点。因此，我们充分发挥场地开阔和高差变化，打造恢宏开阔的大尺度景观。蓝天白云下，钟塔矗立，廊亭簇拥，绿影婆娑，形成动人的天际线；台阶立面的红色花砖，结合中央跌水，如同艺术瀑布，在绿色的草坡中潮水般倾泻而下；白色的带状树池与盛开的草花向人们争先展现纯粹与艳丽。而别致的红砖拼贴立体花柱、青铜小品，闪耀般点缀着竖向空间的层次。

The project is located on one side of the urban rapid corridors in the suburb of Changsha. The entrance design focuses on how to create an impressing symbolic landscape. Therefore, we take full advantage of the open space and altitude differences and create an open and magnificent landscape. Under the blue sky and white clouds stands the tower. The pavilions cluster together and green shadows whirl. All of these form an impressing horizon. The combination of red tiles and central drop water is like the artistic waterfall pouring down in the green grass slopes. White ribbon-like trees pool and the blooming flowers show people their purity and beauty. Besides, the delicate red bricks combined with solid styles and bronze sketches decorate the vertical space levels.

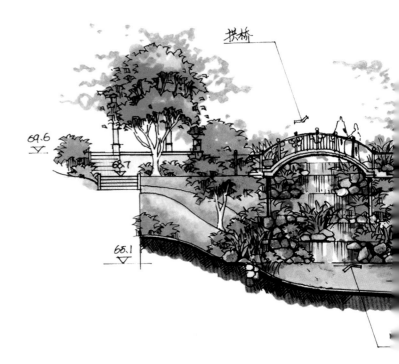

■ 风格独特的售楼处　Sale Department in a unique style

② 体验的景观 —— 仿佛置身公园的湖区
Experiencing Landscape —— Feeling Like in the Lake of the Park

穿过入口景观大门，映入眼帘的是山水一色的两个人工湖。湖面宽广，波光粼粼。湖的旧址为山地下低凹的鱼塘，两处水面有近7米的高差。两者之间以绿岛分隔，岛的两侧河石跌落，植物葱郁。岛与岸边小桥流水，惬意宜人。湖以缓坡入水的形式，结合水生植物，尽显自然景色。蜿蜒的岸线和木质栈桥，离水岸时近时远，或偶尔挑出水面。让人在漫步中赏水亲水，不亦乐乎。

环湖住宅的花园凌驾于湖岸之上，地形与植物的环境营造，保证了私密与公共空间的互不干扰，更为住户提供了居高临下观景的最佳看点。

■ 水中栈道　The plank road in water

■ 清新婉约的上湖景区　Fresh and graceful lakeside landscape area

Passing through the entrance door, we are greeted with two artificial lakes with an integration of the mountain and the lake. With broad surfaces, the lakes are sparkling. The old location of the lake is the hollow fishponds under the mountain. The water levels of the two lakes have a 7meter difference in height. The lakes are separated by a green island. On both sides of the island, there are river stones falling down, lush plants, the bridge and running water, which is very pleasant. The way of drawing water with gentle slopes combined with aquatic plants show the natural scenery. Winding coas-tline and wooden bridges are near or away from the bank as time goes by and sometimes are higher than the surface. People can enjoy appreciating and playing with the water when taking a walk.

The park of the residence surrounded by the lake is over the lake bank. The environment creating of the topography and plants ensures that the private and public space will not disturb each other. Still further, it provides residents the best commanding view of landscape.

■ 静谧的交流小庭院　Quiet Small Linking Courtyards

■ 台地庭院手绘效果图　Manual Effect Drawing of Terraced Courtyards

(3) 生活的景观——精致变化的宅间庭院
Living Landscape
—— Delicate and Changeable Courtyards

公共区域为住户带来了高品质的大环境，而宅间空间作为住户最密切相关的区域，则更讲求安静祥和的生活气息。设计之初优先考虑的是路径的组织。在满足便捷性的前提下，巧妙通过空间平台转换、竖向景观营造，形成亲切宜人的邻里交流、休闲小憩区域。转折的室外楼梯，小巧的水景花池，景墙廊架，让心情找到放松的宣泄点；用植物削弱和柔化建筑轮廓和转角，使得密匝的山墙增添一份清新自然。

Pubic area brings highquality environment. Residential space, as an area closely connected with residents, requires peaceful and harmonious life flavor. The early design gives priority to the organization of roads. With the convenience as a premise, the clever platform exchange and the vertical landscape creating form a pleasant resting area with neighbors communicating with each other. The turning outdoor stairs, small scenic flower pools and corridor walls provide a relaxing place to let out the emotion.

■ 花园庭院手绘效果图　Manual Effect Drawing of Garden Yards

■ 极具米哈斯风格的宅间绿化　Green Landscape Between Houses in Mijas Style

■ 特色的围栏和铺装　Feature Fences and Decoration

结语： 景观是为人而生，当景观的营造切实了人们的生活功能和审美需求，才真正实现了其意义所在。设计师正是本着这一目的，令橘郡米哈斯的环境，带给人们一种对西班牙风情生活的美丽憧憬。

■ 错落的组团入口　Interlocking Entrances of Residence Groups

■ 极具特色的景观挡墙具有很强的景观辨识性　Extremely Characteristic Landscape Retaining Walls with Landscape Identification

别墅景观设计 VILLA LANDSCAPE DESIGN | 大众集团·湖滨花园

■ 全景鸟瞰图　Aerial view of Overall Landscape

■ 景观河道和水岸别墅群　Landscape Rivers and Riverside Villa Groups

岛于墅
地中海风情将在这里上演……

大众集团·湖滨花园

Villa in the Island
Mediterranean style will be presented here

Volkswagen Group · Garden with Lakes

开发商：大众集团公司
项目地址：浙江嘉善
项目规模面积：40万平方
项目类型：大型社区住宅楼盘（公建、洋房、独栋、联排）
景观设计：上海绿茵景园
主创团队：余志国、宋强、刘美丰、崔展
设计及施工时间：2009至今

设计阐述 Design Explanation

场地概况 Location Overview

"水乡泽国银嘉善",精辟的概括了本案所处地域的特点,位于国道0320和沪杭高速公路的嘉善县不仅有便利的交通更有丰富的水资源、浓厚的文化积累等。种种得天独厚的优势都为大众湖滨花园将来成为一个集典雅、高档又独具特色于一体的现代化人文社区打好了坚实的基础。

Location Overview

The saying that Jiashan is a golden region of rivers and lakes summarizes the features of the project location. Located in State Road 0320 and the Shang-hai Hangzhou Expressway, Jiashan County owns convenient traffic and also rich water resources as well as a strong cultural accumulation. All of these advantages lay a solid foundation for Garden with Lakes to become a modern humanity community which is elegant, upscale and unique.

风格定位 Style Orientation

① 拱形的浪漫空间

- **拱门、半拱门、马蹄状的门窗**

  建筑中的圆形拱门及回廊通常采用数个连接或以垂直交接的方式,在走动观赏中,出现延伸般的透视感。墙面常运用半穿凿或全穿凿的方式来塑造室内的景中窗。

  拱形空间为室外环境提供了丰富的"灰"空间。

② 精致的庭院空间

- **水景、植物**

  由于西班牙独特的地理位置,日照充沛,西班牙人喜欢在午后的庭院里喝茶、聊天、休憩。因此庭院是西班牙民宅建筑中不可分割的一部分,通常利用水体和大量的植被来调节庭院和建筑的温度。

③ 纯美的色彩方案

- **黄、蓝、紫和绿**

  向日葵、薰衣草花田,金黄和蓝紫的花卉与绿叶相映衬,形成一种别有情调的色彩组合,十分具有自然的美感。

- **土黄、红褐**

  西班牙戈壁、岩石、泥、沙等天然景观颜色,再辅以西班牙土生植物的深红、靛蓝,加上黄铜,带来一种大地般的浩瀚感觉。

④ 不修边幅的线条

- **自然、古朴、浑圆**

  线条是构造形态的基础,因而在景观营建中是很重要的设计元素,西班牙很多房屋或景观中构筑的线条不是直来直去的,显得比较自然,因而无论是景观还是建筑,都形成一种浑圆的造型。

■ 独具西班牙特色的棕榈大道　Palm Avenue in Spanish Style

■ 入口鸟瞰之一　One of Entrance Aerial views

■ 西班牙风情的亲水平台　Waterfront Platform in Spanish Style

⑤ 独特的装饰方式

- **碎瓷、陶砖、拼贴**

  环境注重采用低彩度、线条简单且修边浑圆的木质材料，地面则多铺赤陶或石板。马赛克镶嵌、拼贴算较为华丽的装饰。主要利用小石子、瓷砖、贝类、玻璃片、玻璃珠等素材，切割后再进行创意组合。独特的锻打铁艺家具，也是西班牙风格的特色之一。同时，花园中爬藤类植物是常见植物，小巧可爱的绿色盆栽也常看见。

■ 设计代入 Design Empathy

- **造园手法**

  在设计中，整体形状和布局至关重要，通过这两者把各种装饰元素组合在一起，从而影响整体的观赏效果。为了达到这种既古典又时尚的效果，在材料的选择上十分重要。同时与环境进行有机的结合，营造丰富的室外灰空间和观景的场所，营造出幽雅浪漫的西班牙风情。

- **代入元素I**

  庭院 （Patio帕提欧）
  围和较强的庭院空间
  引入农业灌溉的水景方式
  修剪整齐的植物
  令人惊奇的空间变化

- **代入元素II**

  色彩和图案
  小品和构筑

  景观色彩上做一些与西班牙风情相呼应的色彩。泥绿色、海蓝色、锈红色和土黄色的背景幕墙，土色与褐红色的陶罐里常种植粉红和红色的花草。在小品和构筑物方面注重色彩图案的艺术性和连贯性，可结合毕加索、高迪等大师的作品设计成雕塑和景观柱。同时，也要注意运用凉棚等营造幽静阴凉的环境，座椅上摆放一些具有传统色彩的软垫和垫木以取得柔和的效果。此外，木材、藤条和金属则是最受欢迎的装饰材料，陶罐也是必不可少的装饰元素。

■ 萨拉戈萨风情岛别墅入户口效果图
Effect Drawing of Household Entrance of Island Villas with a Zaragoza Flavor

- **代入元素III**

  铺地和材料

  地面材料可以选用未经打磨的粗糙石板，或是乡村风格的瓷砖。沙砾适用于一些比较休闲的场合，用来填充边沿和走动较少的地方。大块的鹅卵石可用来拼成曲折的线条和装饰小径路面。一般庭院和内院都铺有陶瓷砖、机制砖或石块。

  植物

  本案中有着独特的自然环境，在基地附近是主要以桉树、栾树、刺桐为主的混生林带，我们主张尽力保留。但是西班牙氛围较弱，那么我们还需要通过植物配置来增强西班牙风格。

  主要运用的具有西班牙风格的植物有龙舌兰、橄榄树和棕榈树，还有针叶类植物。

  此外，石榴、玫瑰、无花果和葡萄也都是必不可少的。

■ 景观意向剖面 Reference Profile of Landscape

■ 融合西班牙式塔楼风情的社区入口　Community Entrance in a Spanish Tower Style

古堡情趣、闲散文化、浪漫动人的景观细节，热情奔放的艺术情怀

It is designed to have castle fun, restful culture, romantic landscape details and passionate artistic sentiments.

■ 华丽大气的南入口鸟瞰和透视效果
　Splendid Aerial View and Perspective Effect of South Entrance

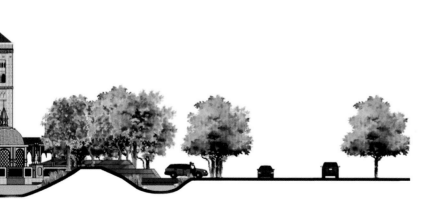

■ 充满异域风情的商业街一角　A Corner of Business Street with an Exotic Flavor

花园洋房景观设计
Western-style House Landscape Design

专属空间的魅力

花园洋房做为介于普通公寓与别墅之间的一种建筑类型，是西洋文明、生活方式与中国文化融合的新兴居住类型，其凭借林水相依的自然景观、丰富的休闲娱乐配套设施，以私家专属的丰富情趣空间为独特魅力。

比肩别墅的一流居住环境

花园洋房作为低密度、社区绿化率高，私密度强，拥有大的入户花园、露台，能够给人带来舒适环境的多层建筑。最打动人心的，在于其并不昂贵的价格，与一流的居住环境。本刊以蜀山西镇、戛纳印象等为例，阐述设计师在花园洋房景观设计中的独特理念。

花园洋房景观设计
Western-style House Landscape Design

82	90	94	102	108

- 维丰·蓝湖熙岸 Weifeng, Bank of the Blue Lake
- 蜀山栖镇 Shushan Qi Town
- 戛纳印象 Cannes Impression
- 龙湖·三千城 Longhu, three-thousand City
- 龙湖·弗莱明戈 Longhu, Flegory

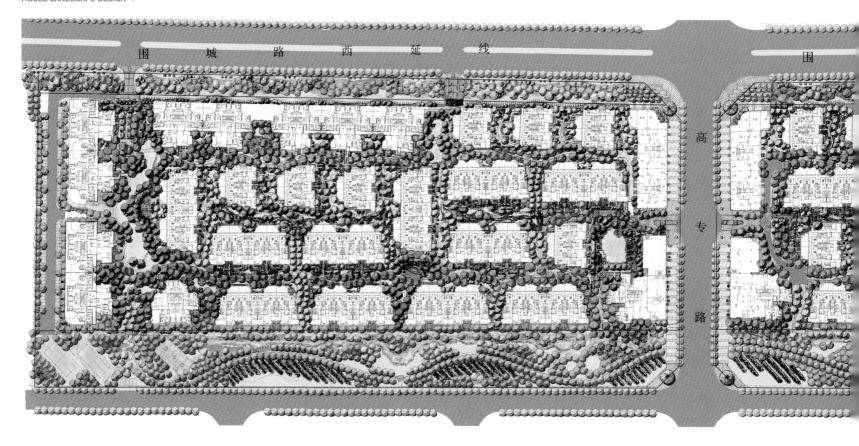

邂逅花园慢时光
龙湖·弗莱明戈

Enjoying Leisure Time in the Garden
Longhu Flegory

项目开发商：成都西玺置业有限公司、成都西祥置业有限公司

项目地址：成都市郫县

用地面积：70000平米

施工图设计公司：成都绿茵景园 设计二所

主创团队：温丽梅、马卉、刘丽红、雷冬、粟凡粒

发展阶段：2009年

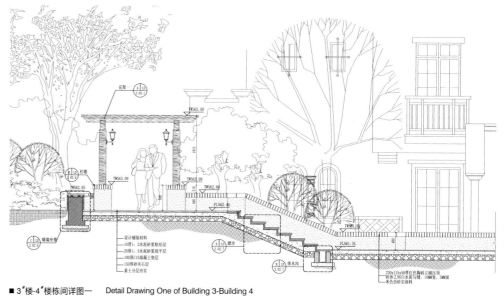

■ 3#楼-4#楼栋间详图一　Detail Drawing One of Building 3-Building 4

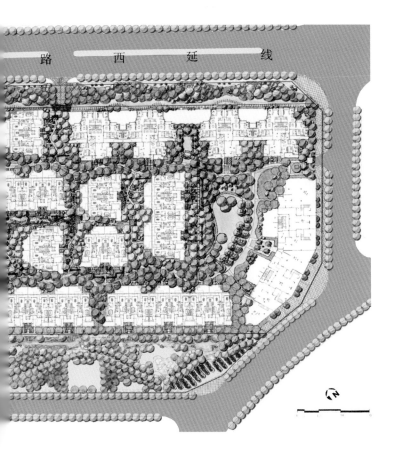

■ 极富西班牙风情的花灌与小径　Flowering Shrub and Roads with a Spanish Flavor

设计思路 Design Idea

总体景观方案的深化设计是在延续总体概念方案的基础上，依据"西班牙坡地小镇"的景观构思，至始至终贯彻了以下设计原则：

① 以别墅景观设计的手法设计公寓区景观：

别墅景观设计与公寓景观设计相比，更注重空间的归属感，我们在设计中运用该手法，目的在于提升本项目住户对于场地的拥有感，实现景观价值的外溢；

② 以庭院的尺度塑造公共空间：

本项目密度较高，景观空间狭小，只有采取小中见大的手法。将原规划较为通直的消防通道改为蜿蜒曲折的园路，并减小路幅，使回家的路更加亲切宜人，并被花丛簇拥；同时，将庭院作为主要的设计元素，以室内化的细节处理室外空间，彰显了西班牙的景观特色。

The detailed design of the overall landscape plan is based on following the overall concept plan. The design carries out the following planning principles from the beginning to the end according to the landscape design con-ception of "Spanish Slope Town".

1. Design the apartment area landscape in a way of designing the villa area landscape:

Compared with the apartment area landscape design, the villa area landscape design emphasizes the returning sense of space. We design the apartment area landscape in this way in order to elevate the Owners owning sense of site in this area and to realize the extension of the landscape value.

2. Create the public space with the size of the court:

The project has a higher residence density and the landscape space is limited. Therefore, the dealing way of presenting the open space in a limited area is taken. The straight fire fighting access in the original planning is changed into a winding garden road. Besides, the breadth of the road is reduced to make the way home more pleasant and the road is surrounded by flowers. At the same time, the courtyard is designed as a main element. Dealing the outdoor space in indoor details presents Spanish landscape features.

■ 红砖路边的植被群落及休憩空间　Vegetation Groups and Rest Space on the Side of red Brick Road

■ 红砖黄墙的西班牙式种植池
Planting Pool with red Bricks and Yellow Walls in Spanish Style

③ 处理好公共空间、半公共空间和私有空间的关系：

尽可能地将半公共空间化为私家庭院的功能使用，以增加空间的使用性。

④ 景观设计考虑建设的成本，控制软硬景的比例：

尽可能的减少大面积铺装和水景，体现景观设计的经济性和合理性。植物景观设计密切结合项目的主题，在考虑适地适树的同时，也选用了特色树种。

3. Handle the relationship among public space, semi-public space and private space：

The semi-public space should be used as private space as much as possible to increase the use of space.

4. The landscape design should take the construction cost into consideration：

The ratio of soft and hard landscape should be controlled. And large area of decorating and water landscape should be avoided to present the economics and reasonableness. Plants landscape should be connected closely with the project's theme and feature tree species should be chosen based on the consideration of appropriate trees and appropriate area.

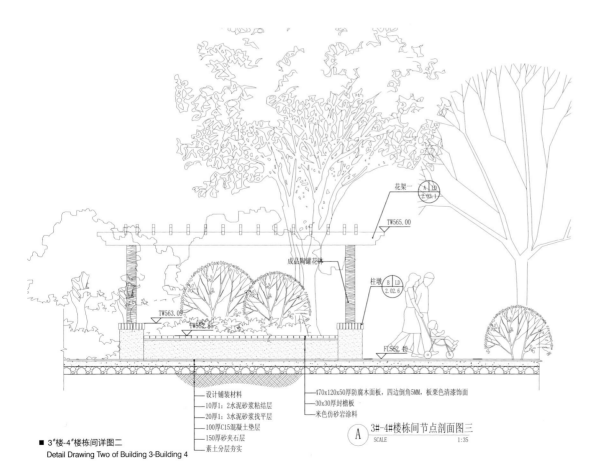

■ 3#楼-4#楼栋间详图二
Detail Drawing Two of Building 3-Building 4

卵石，φ50-200，杂色，黑色居多，密拼，底层粘接固定
20厚1:2水泥砂浆结层
100厚C15混凝土保护层
8厚安能膨润土防水毯
50厚粘土层
80厚细砂垫层
素土分层夯实
排水防水做法（详建施）
地下室顶板结构层（详建施）

(φ400-1000)雅安河石
水泥砂浆灌缝（添加3%防水剂）

设计铺装材料
10厚1:2水泥砂浆粘结层
20厚1:3水泥砂浆找平层
100厚C15混凝土垫层
150厚砂夹石层
素土分层夯实
排水防水做法（详建施）
地下室顶板结构层（详建施）

A/LD 入户木桥
1.09.1

FL565.80
WL565.60
BP565.30

SL564.55

汀步 C/SD 1.02.2
M5水泥砂浆砌MU10标砖
排水暗沟 A/SD 4.02

■ 阳光溪谷二剖面图 Two Profile Drawings of Sunshine Valley

注：卵石之间密拼，间隙较大处勾缝；施工时粘结卵石的规格从池底到池边逐渐变大；

■ 白色拱形门廊与蓝色的游泳池　White vaulted porch and blue swimming pool

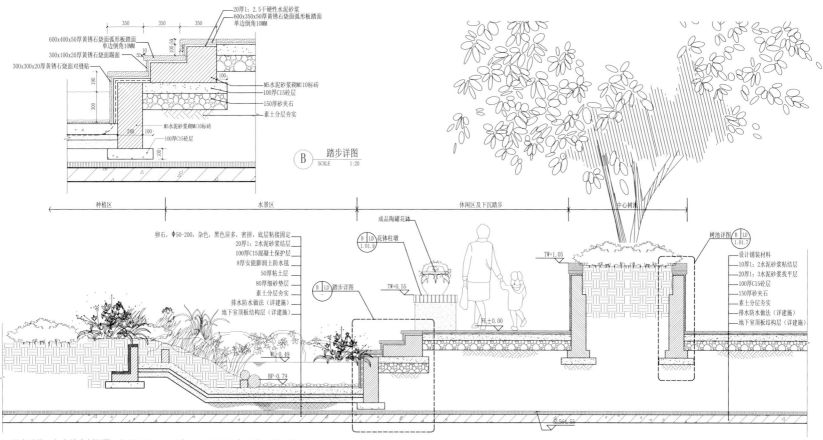

■ 阳光溪谷—入水踏步剖面图　Profile Drawing of Sunshine Valley-Step Road into Water

花园洋房景观设计 | 龙湖·弗莱明戈

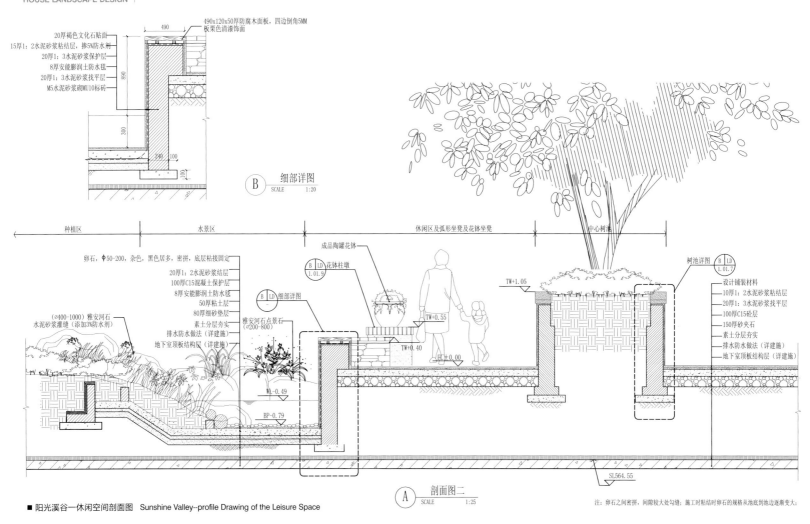

B 细部详图 SCALE 1:20

A 剖面图二 SCALE 1:25

■ 阳光溪谷—休闲空间剖面图　Sunshine Valley--profile Drawing of the Leisure Space

注：卵石之间密拼，同隙较大处勾缝；施工时粘结时卵石的规格从池底到池边逐渐变大；

■ 窗外花团锦簇　A Cluster of Flowers out of the Window

■ 阳光照耀下的后花园　Back Garden with Shining Sunshine

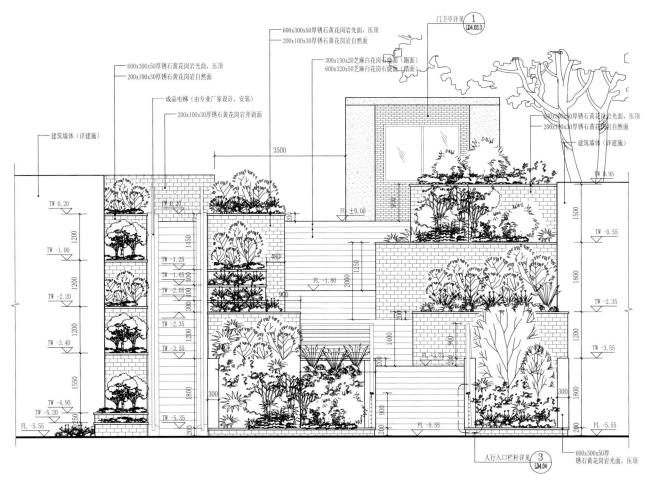

城市上空的森林

龙湖·三千城

Forest up the City
Longhu, Three-thousand City

项目开发商： 成都龙湖同晋置业有限公司
项目地点： 成都成华区建设路二环路东2段3号
项目面积： 规划用地7.15万平方米
景观面积： 约6万平方米
项目类型： 现代尊贵风格居住区
设计单位： 成都绿茵景园 设计一所
主　创： 潘旭、潘迪
团　队： 粟凡粒、张彬、姚抒雅、陈浩然
设计时间： 2007年至今

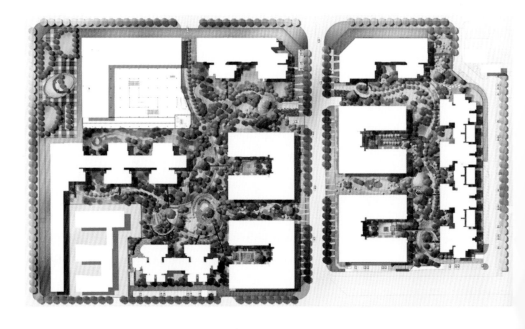

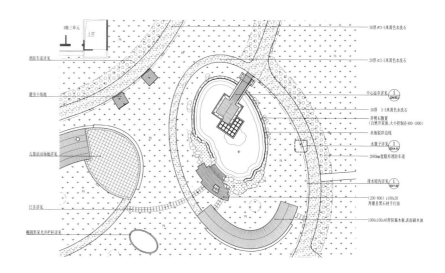

设计说明 Design Explanation

"三千城"项目景观设计旨在与建筑"现代"的整体设计概念相匹配,凸显出小区的尊贵感。创造出森林中绿岛的景观效果,给人以深刻印象。小区人行主要入口层层台阶配以错落式的种植池并采用自动扶梯使台阶的生硬感消失,也使得入口显得更生动,精致,提升了整个小区的品质感,凸显其尊贵的气质。

项目的景观设计理念就是要营造森林中的绿岛(一种现代,生态又不失尊贵)的一种景观。此项目的设计理念就是要在布局中采用主导元素,包括水体、小径、软景、墙体、景亭、活动场地,重新演绎这些元素之间的关系,并通过这种做法营造一种现代与自然相结合的景观。

园区中心设置自然式的水景配以木质的特色景观方亭,营造出一种自然又不失精致感的现代景观。旁边椭圆形的儿童活动场地,配以丰富的植物,为儿童提供活动场地的同时也为小区景观增色不少。在小区局部还对架空层加以利用,为业主提供更多的活动体验空间,即便是在下雨的天气,也不影响活动。小区在植物配置上以乔灌木、草木花卉等组成软质景观,为户主提供自然清新的林中行路体验,让户主在繁忙生活中重新发现自然之美。观赏、散步、游憩的主要道路,行在其中既可观赏景色又可体验幽静休闲的气氛。最后利用小区主要道路把所有元素串联在一起,达到既美观又能满足感官和功能上的需求。

■ 每一个组团都有其独特的标识 Every group has its unique sign

■ 组团中央别致的休闲平台，在矮墙与花径的映衬下，休闲的氛围油然而生
There is a leisurely platform in the center of the group and leisure atmosphere is produced against the parapet and road with flowers

■ 被水体环绕的观景亭
Sightseeing pavilion surrounded by water

■ 观景小径畔贴心的休闲座椅
Sweet recreational chairs along the sightseeing roads

　　小区中建筑设计有四栋花园洋房，U形的建筑设计平面，为还原洋房的业主提供了一块有别于园区中大环境的景观空间。利用地面石材铺设方式的变换与小水景材质的对比配以丰富的植物，营造出一种精致、尊贵的洋房景观。让花园洋房的业主在体验完园区自然与现代相结合的风景之后，再来体验洋房区精致尊贵的景观感受。

　　总体景观为业主营造了出了一种生活中的人文之美，环境中的舒适之美与活动中的动感之美。

府河之畔，戛纳之念
戛纳印象、戛纳湾、戛纳滨江

Bank of the River, Dream of Cannes
Cannes Impression, Cannes Bay, Cannes River

项目地址：四川成都华阳滨河路二段
开发商：四川同森集团
项目类型：商业、住宅、酒店
景观面积：戛纳印象163亩 戛纳湾70亩，戛纳滨江108亩
合作单位：美国ESD、贝尔高林
景观设计：成都绿茵景园 设计三所
施工图主创团队：肖浩波、潘强、曾健、潘宇、陈浩然
设计时间：2006至今

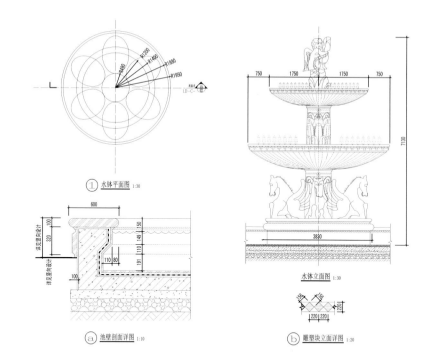

① 水钵平面图 1:30
水钵立面图 1:30
ⓐ 池壁剖面详图 1:10
ⓑ 雕塑块立面详图 1:20

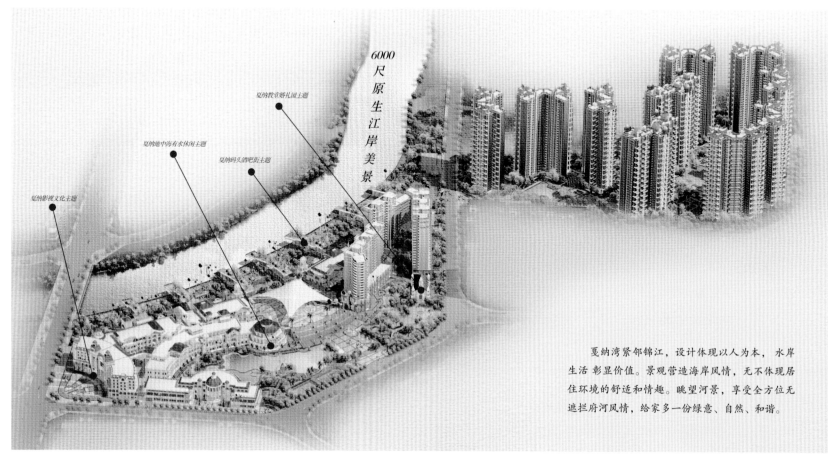

戛纳影视文化主题
戛纳码头酒吧街主题
戛纳地中海有水休闲主题
戛纳教堂婚礼园主题
6000尺原生江岸美景

戛纳湾紧邻锦江，设计体现以人为本，水岸生活彰显价值。景观营造海岸风情，无不体现居住环境的舒适和情趣。眺望河景，享受全方位无遮拦府河风情，给家多一份绿意、自然、和谐。

■ 戛纳湾畔夜景 Night Scene at Cannes Bay

总体概述 Overall Description

在成都，府南河养育着平原儿女，成都以南，府南河流经的地方，两岸人民更是依河而居。有那么一个地方，府河湾畔，戛纳人家，成就了千千万万的戛纳之念。

地处华阳滨河路二段，由同森集团开发的戛纳系列楼盘，集商业、娱乐、住宅、酒店等于一体。戛纳印象以商业娱乐为主，戛纳湾以商业酒店休闲为主，而戛纳滨江则以住宅为主。

戛纳印象城南"休闲娱乐文化"中心，含婚庆文化产业、电影文化基地、风情商业港、餐饮休闲娱乐、特色商业购物街、五星级风情酒店、大型超市等七大主题。项目风格为欧式风格，地中海风情。浪漫地中海，更是设有蜡像馆、水疗馆、游乐园、酒吧、水幕电影等各种休闲娱乐项目。项目在设计、施工、商业环境营造等方面精益求精，致力于打造"立足城南、面向成都"的独一无二的商业文化基地，"戛纳印象"将成为成都最具魅力、最有价值的特色商业中心。

戛纳湾紧邻锦江，设计体现以人为本，"水岸生活"彰显价值。景观营造海岸风情，无不体现居住环境的舒适和情趣。眺望河景，享受全方位无遮拦府河风情，给家多一份绿意、自然、和谐。

戛纳滨江建筑依江而建，近2000米锦江江景尽收眼底。纯住宅景观，并坐拥全部戛纳配套，旨在打造高品质居家环境，还住宅的静谧本原。建筑景观相结合，将建筑与景观相互融合，体现景观独有特色：低密度、架空层、五星级大堂入户、人车分流等。整个小区分组团，每个组团各具特色。

■ 每一个细节的打造都独具匠心
Every detail is created with uniqueness.

■ 沿河步行街区，享府河自然风光　Riverside pedestrian street area, pleasant the natural sight of Fu River

Overall Description

In Chengdu, Funan River provides water for people in plain. In the south of Chengdu where Funan River flows through, citizens on the bank live along the river. There is a place at the bank of Fu River. The community of Cannes realizes many people's Cannes dream.

Cannes Series Building is located in the second section of Binhe Road and developed by Tongsen Group. It is an integration of business, entertainment, residence and hotel. Cannes Impression deals with business entertainment, Cannes Bay with business hotel and leisure and Cannes River with residence.

There is a "Leisure, Entertainment and Culture" center in the south of Cannes Impression, with seven themes of wedding celebration culture industry, film culture base, elegant commercial port, catering entertainment, featured commercial shopping street, elegant five-star hotel and large supermarkets. The subject is of European style with a Mediterranean flavor. In the romantic Mediterranean, there are various choices of entertainment, including a waxwork museum, water spas, amusement ground, the Ferris wheel, bars and films with a water curtain. The subject makes perfection more perfect in the design, construction, and the creating of commercial environment, aiming to create a unique commercial culture base which "bases itself upon the south of the city and faces Chengdu". Cannes Impression will be the most charming and valuable featured commercial center in Chengdu.

Cannes Bay is besides Jin River. The design presents the idea of people oriented and "the life along the river" reveals its value. The landscape creates a coastline flavor and shows the comfort and interest of the living environment. Looking far at the river landscape, you can enjoy the all-round landscape of Fu River without blocking. You can give your home a sense of greening, nature and harmony.

The architecture of Cannes River is constructed along the river with a landscape of 2000-meter Jin River. There is a pure residential landscape Cannes River with all Cannes suites. It is aimed at creating a high-quality living environment and restoring the quiet of residence. The combination of the architecture and landscape integrates the architecture and the landscape and presents the uniqueness of the landscape.

Features: Low density, Stilt floor, entrance of the five-star hall and the separation of pedestrians and vehicles. The community is divided into divisions, each with its own features.

■ 河边休闲茶座　Leisurely tea house along the river bank

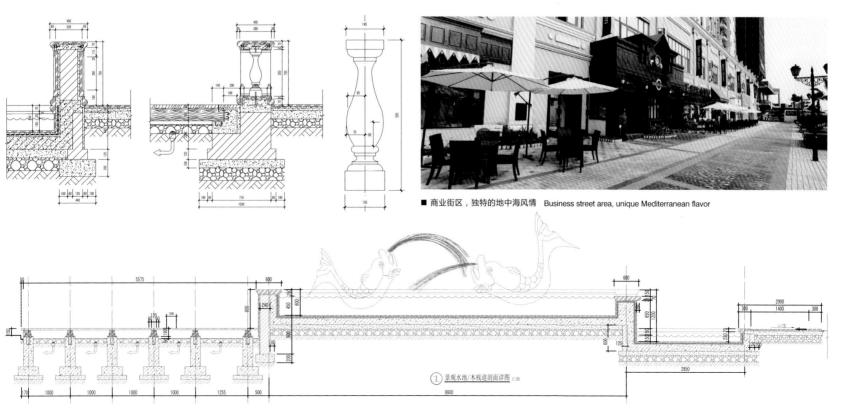

■ 商业街区，独特的地中海风情　Business street area, unique Mediterranean flavor

① 景观水池/木栈道剖面详图 1:20

■ 不经意的一个角落都很美 A corner by a casual glance is very beautiful.

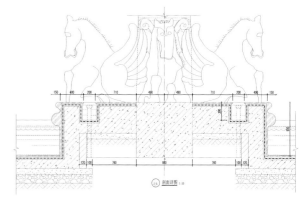

■ 退台造型植物景观 Plants landscape in a form of terraced roof

■ 婚庆广场户外草坪 Outdoor grassland of Wedding Square

■ 浪漫神圣的婚礼广场 Romantic and holy Wedding Square

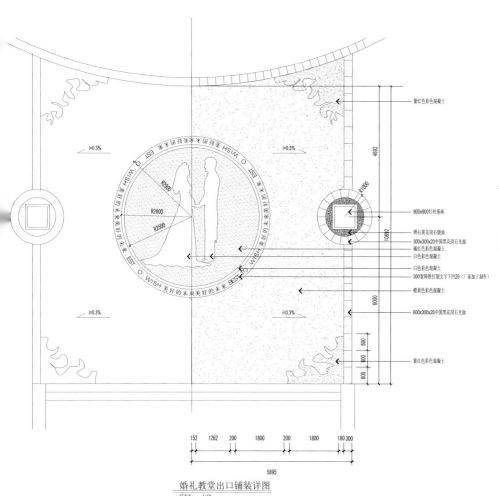

婚礼教堂出口铺装详图
SCALE: 1:50

特色铺装详图（三）
SCALE: 1:30

景观分区 Landscape Divisions

第一组团：江锦 — 纯水纯景
The First Division: Pure Water Pure Sight

江锦，拥有领衔戛纳滨江的丰富江景，与15万m²戛纳印象一水之隔，近繁华而远尘嚣。尽揽100亩园林绿意和浩瀚江景，一线临江的亲水优势，江锦的王者之气已经蔚然成型，令国际城南日益稀缺的水上人居体验再度升级。

There is abundant river scenery in leading Cannes Rivers in Jiangjin. Cannes Impressions is separated from Jiangjin by a river and is close to prosperity but far from bustle. With a green garden of 100 mu, vast river scenery and the advantage of being close to water, Jiangjin has achieved the dominant momentum. Thus, the rare residence on the water in the south of the international city has been upgraded.

第二组团：尊贵湾区品质生活
The Second Division: Honorable and High-quality Life on the Bay

坐拥7000平米景观园林绿意，6000尺锦江秀色，800米河堤休闲栈道，沿岸异域风格的欧式建筑，海滨风情餐饮酒吧街，8000平米的"爱琴海"，地中海风情沙滩泳池，阳光网球场……占据稀缺景观资源，以自身的完美品质和世界同步。

It has a green garden of 7000 square meters, 6000-chi-long Jinjiang River scenery and a leisure road along the 800-meter-long river. Along the bank, there is exotic European architecture, catering street in a seaside style, "Aegean Sea" of 8000 square meters, a beach swimming pool in the style of Mediterranean Sea and the tennis court with sunshine. It occupies the rare resources and keeps up with the world with its perfect quality.

第三组团：原乡别墅园林美景
The Third Division: Beautiful Scenery of Rural Villas

500m²枫木林，20米林荫道只是你归家路上的美丽风景。富含浓郁法国浪漫气息的戛纳湾生活从此唯君尊享！

The maple forest of 500 square meters and the 20-meter-long boulevard are only the beautiful scenery on your way home. From now on, you can enjoy an honorable life in Cannes Bay which has a rich flavor of French romance.

■ 商业酒店一角　A Corner of the Business Hotel

■ 商业绿化空间　Business Green Space

花园洋房景观设计
WESTERN-STYLE HOUSE LANDSCAPE DESIGN

蜀山栖镇

■ 鸟瞰图
Aerial View Drawing

山水之间
独享精致生活的叠园叠院

蜀山栖镇

Between the Mountain and the River
Enjoying an Exquisite Life in the Courtyard

Shushan Qi Town

开发商：成都传媒集团蜀山投资有限公司
地点：成都崇州市街子镇
规划用地面积：320亩
景观设计：成都绿茵景园 设计二所
主创：韩茜
参与方案设计：张洪源、刘鸠鸠
硬景施工图团队：马卉、董玲、袁怡、刘丽红
专业组：粟凡利、潘宇、雷冬
发展阶段：2009年
完成阶段：已完成一、二期园区景观及滨江带商业街景观

设计说明 Design Explanation

诗意的栖居,从山水说起。

蜀山栖镇拥有得天独厚的自然山水资源,依山傍水,在设计之初就具备无可复制的天然,即有自然水墨画的神韵。"大山水"的天然优势使我们更加关注自然与人之间的关系,道法自然,运用各种设计元素,如地形、植物等来丰富空间层次,达到自然的回归。所以设计顺势而为,借山之秀气、水之灵气,将"自然而然"的主题和"观山、听水、慢生活"的理念贯穿设计始末。

设计从山水视觉的引入出发,充分利用街子的山水特色,赋予景观环境亲切宜人的感召力,通过景观别墅与底层度假公寓结合、外部味江河河滨绿地、售楼部滨河示范区、与小区内部水景、组团景观的完美融合、互动与演绎,以舒适宜人的山水景观和丰富的邻里空间组团打造出"外部山水大自然,内部活水小自然"的新型慢生活社区。

The poetic residence is introduced from the mountains and the rivers.

Natural resources of mountains and rivers are unique in Shushan Qi Town. It is located against the mountains and by the rivers. It had unique natural element at the beginning of the design, that is, the charm of nature wash painting. The natural advantage of "Great Mountain and River" attracts more of our attention to the thinking of the distance between nature and human being and nature rules. Various elements are used, such as the terrain and pla-nts to enrich the space levels. In this way, nature return is achieved. Therefore, it is designed taking advantage of the elegance of the mountain and the liveliness of the water. The theme of "Naturally" and the idea of "Leisure Life with Appreciating the Mountain and the River" run through the design.

The design begins with the visual introduction. It takes full advantage of mountain and river characteristics in Jiezi to give the landscape emotional appeal. There is a combination of landscape villa with the lower holiday apartment, external beach Greenland of Weijiang River, river demonstration area in sale department, internal water landscape in the community, and the perfect integrating, exchange and presence of group landscape. Comfortable and pleasant mountain and water landscape and abundant neighboring space group create a new type of leisure life with "External Nature with Mountains and River and Inte-rnal Environment with Lively Streams".

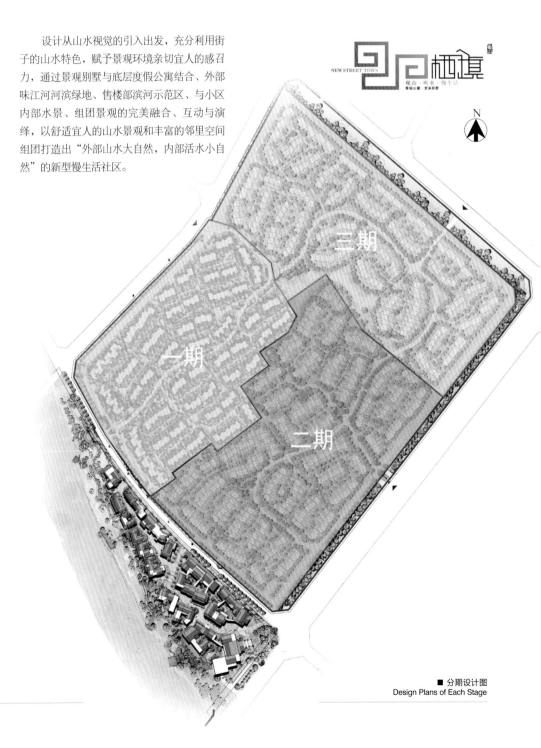

■ 分期设计图
Design Plans of Each Stage

花园洋房景观设计 / WESTERN-STYLE HOUSE LANDSCAPE DESIGN

蜀山栖镇

图例：

- 01. 景观绿岛
- 02. 新印象街子广场
- 03. 景观台阶
- 04. 会所酒店主题景观
- 05. 特色山石叠水
- 06. 生态停车场
- 07. 水街商业风情街
- 08. 水巷子
- 09. 水街绿岛
- 10. 门庭景观
- 11. 草坪高尔夫
- 12. 黄葛树集会广场
- 13. 诗意走廊
- 14. 林荫步道
- 15. 江村码头
- 16. 中庭花园
- 17. 林荫休闲平台
- 18. 亲水平台
- 19. 尊享空间
- 20. 后院观景花园
- 21. 亲水平台
- 22. 竹笠卵石箱
- 23. 观景亭
- 24. 曲院水街
- 25. 点景大树
- 26. 观景后花园
- 27. 观景台
- 28. 游步道
- 29. 节点小广场
- 30. 景观水面
- 31. 景观墙
- 32. 瀑布
- 33. 鹭鸟
- 34. 溪流跌水
- 35. 公园入口
- 36. 疏林草坪

■ 滨水园林式商业水街平面总图
Overall Plane Layout of Business Streets in Waterfront Garden Style

设计秉承这"传统水墨美学为体，现代简约为用"造园理念。溪谷穿行其间，一步一景色变换，使每栋住宅都有自己的专属景观，前庭后院，户户皆有独到风景。看见风景的房间，远\中\近景如画舒展，青山入目，碧绿入窗，独享自己的闲居天地。园内景观一切归于自然而然的流畅，返璞归真的亲切……

The design follows the planning idea of "with a basis of traditional ink aesthetics, with a style of modernism and simplicity". Streams run through the houses and the scenery changes as people move. Therefore, every house has its exclusive landscape. Special scene can be seen from each house, no matter in the front yard or the backyard. In the room with scenery, the landscape far away, less far away and the landscape in front of us come into sight as if a picture were unfolded. People can see the green mountains and enjoy their private space. The landscape in the garden is designed by natural flow and pure pleasure.

■ 精致中式会所入口广场 Exquisite Entrance Square in Chinese Style

■ 融入自然的商业水街　Waterfront Business Streets in Combination with Nature

创作的过程犹如营造水墨画般的写意，活水泼墨，画笔轻匀，点缀上树木、花草和景石，走转而成：

一山一自然、一江一风光、一街一浪漫、一路一心情、

一园一风景、一溪一灵动、一巷一故事、一院一悠然、一广一画境；

九级递进空间层次，彰显中式院落精神、隐居山林溪谷的生活美学。

The process of creating is like painting. Splash ink, paint with the brush, decorate the painting with trees, flowers, grass and landscape stones. Finally the painting is finished with a natural mountain, a scenic river, a romantic street, a pleasant feeling, a beautiful garden, a lively stream, an alley with a story, a yard with leisure and a house with scenery. A five-stage progressing administrative levels reveal the spirit of Chinese yard and the aesthetics living in the mountain forest.

■ 毗邻味江河的商业临水区　Waterfront Business Area Next to Weijiang River

■ 一期景观大道标准段立面图　Elevation drawing of first-stage landscape avenue standard section

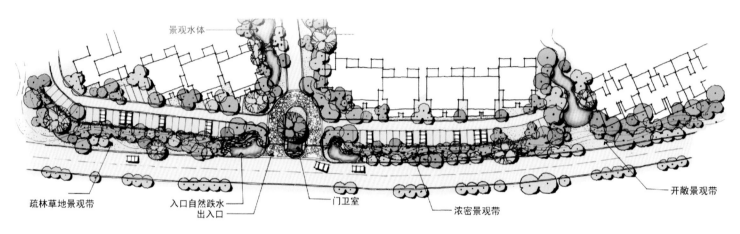

造园追寻一种境界，一种心境。从视觉、听觉、嗅觉等感官体验出发，具有节奏层次的组织景观，景观环境不仅可观，更可以停留；亲水平台等小空间的设计，使居住者与山更近，与水更近。

脚步，因风景而慢。

The construction seeks for another effect, the mental state. Starting from the sense of vision, hearing and smelling experiencing, people can not only see, but also maintain the landscape with a rhymed level in mind. The designs of small space such as platform close to water make the residents closer to the mountains and rivers.

You will walk slower to appreciating the beautiful scenery.

■ 一期样板体验区　Experiencing Area of first-stage Sample

■ 中心湖尊贵岛屿　Noble islands of Central Lake

"自然而然 水墨风景"

山水之间
自然之境界
观之安然……
听之畅然……
处之释然……
入之悠然……
悟之豁然……

二期造园三大特色：
Three Features of the Second Stage

对"自然而然"整体风格的延续与升级，二期景观设计更多融合传统中式园林的空间层次，打造山水之间独享精致生活的叠园叠院：

Following and upgrading the overall style of "naturally", the second-stage landscape is designed mostly in combination with the space levels of traditional Chinese Garden and created with private courtyards to lead an exquisite life.

There is a garden in a garden, a yard in a yard and scenery in scenery ……

山水之间独享精致生活的叠园叠院：
园中园、院中院、景中景……

七彩梦溪：巨木翳然，水出峡中，停蓄杳缈，弥弥之一伤者，目之梦溪。溪之上等为邱，千木之花缘鱼者，百花维叠处。腹堆而庐其间者，蜀之拓处。其西屑于花竹之间，蜀之所频者阁侨于阡陌，克轩处，轩之嵌，巨木百千咏其主者，花堆之阁处……

沈括《自志》

采菊东篱的田园生活
维丰·蓝湖熙岸

A Leisurely Rural Life
Weifeng, Bank of the Blue Lake

项目类型：现代中式风格居住区
项目面积：20万平方米
设计单位：重庆蓝调国际（绿茵景园集团公司）设计一所
项目地点：重庆长寿区
委托单位：重庆维丰金源置业有限公司
主　创：任刚
团　队：毕林枫、樊菁、黄茜、陈果、李海坪
设计时间：2009年

■ 百花叠溪——梦溪覆锦花叠溪　Stream with Clustering Flowers —— A Fantastic Stream Surrounded by Clustering and Flourishing Flowers.

■ 花径林香——用手感悟自然之美
Flower Road with Sweet Flavor of Forest —— Appreciating the Beauty of Nature with Hands

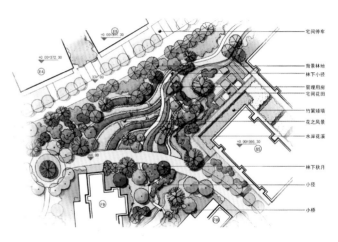

■ 花径林香放大平面
Enlarged Plan of Flower Road with Sweet Flavor of Forest

■ 井源通幽——古韵悠然起新趣
The Quiet And Secluded Well —— An Ancient Well with Fresh Pleasure

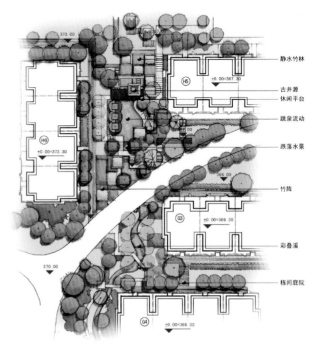

■ 井源通幽放大平面
Enlarged Plan of the Quiet and Secluded Well

设计简介 Design Introduction

项目景观面积约20万平方米，东西宽约560米，南北深约530米。整个场地地形较为复杂，呈西北方高、东南方低的形态，场地中间有一条冲沟，形成局部的沟谷景观。在场地内部，高差也比较复杂，形成非单一坡向的地形。设计师运用现有的湖居文化将景观环境设计与"以人为本，以人与自然的和谐共生"协调融合，相映成趣。设计风格上依托于长寿湖景区，以其"岛湖风光"、"长寿文化"、"乡土文化"为主要景观特色，将打造集休闲养生、旅游、度假、居住为一体的地域性现代中式风格社区。快节奏生活产生的快餐文化显得苍白无力，不能满足人们的精神需求，人们希望借一些自然景物如梅、竹等为生活赋予一定的文化内涵，表达对清雅含蓄、端庄丰华的东方式精神境界的追求。现代中式风格不是纯粹的元素堆砌，而是通过对传统文化的认识，将现代元素和传统元素结合在一起，以现代人的审美需求来打造富有传统韵味的事物，让传统艺术的脉络传承下去。

The project landscape covers an area of 200 thousand Square meters, a breadth of 560 meters from east to west and a depth of 530 meters from south to north. The site's terrain is complicated. It is higher in the northwest and lower in the southeast. It has a gully in the middle creating a partial gully landscape. Inside the site, there is a complicated altitude difference which forms a terrain of nonsingle slope direction. The designer uses the present culture of living by the lake to combine the landscape design with the principles of "people oriented" and "living in harmony with nature". The design and the principles are beautifully coordinated and add fun to each other. The design style is based on Longevity Lake Resort and is featured by "Island Lake Scenery", "Longevity Culture" and "Local Culture". It aims at constructing a regional modern community in Chinese style integrating leisure, health keeping, tourism, going vocation and living. The fast food culture that originates from fast-pace life is of little help and can not satisfy people's spirit need. People hope that they can make use of some nature landscape such as plum blossoms and bamboos etc. to bestow certain culture connotation on life. In this way, they can express their seeking for the elegant, reserved, dignified and luxuriant Oriental Spirit. Modern Chinese style is not a sheer stack of elements, but a combination of modern elements and traditional elements through the knowledge of traditional culture. Besides, modern Chinese style processes things with rich traditional flavor based on modern people's aesthetic need to pass on traditional culture to next generation.

■ 烟雨画屏——梦幻飘渺的诗情画意
Painted Screen in the Misty Rain——A Misty Dream with Poetic and Artistic Conception

在设计原则上强调景观环境的整体性、度假性；自然生态、远离都市烦嚣，憧憬山水田园的家居理想，将作为景观设计的目标。同时将社区功能引入小区规划，满足人们的社交要求。设计突出主题文化，具有先进性、独特性、均好性、人文性、舒适性、实用性、经济性和规范性；以人为本：坚持以人为本原则，体现对住户的关怀，创造多样性的生活度假空间，讲求空间的多层次性、连续性和人性化设计；生态优先：尊重自然地形地貌，强调人与自然、建筑与自然、人工环境与生态环境的和谐，充分利用地形营造居住环境，营造舒适、休闲的居住度假空间；经济实用、美观艺术：利用小区原有地形、地貌、水体、植被等自然、人文条件，合理造景。注重景观效果与控制造价相结合。植物营造注重地域性。

在功能活动上强调趣味结合，在总体设计中，加入了农场、苗圃等趣味概念，在植物设计上也做总体考虑，在不破坏整体性的原则下，加入更多的景观元素，以达到崇尚自然情趣的景观效果。

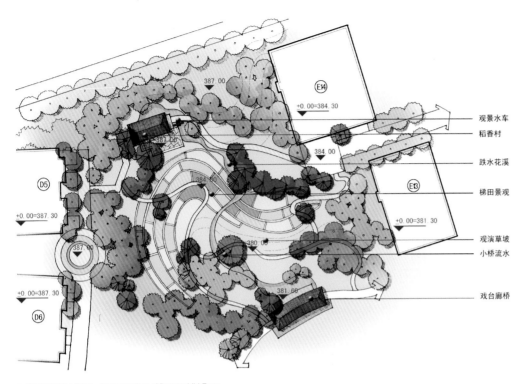

■ 亲子农场放大平面　Enlarged Plan of Parent-child Farm

观景水车
稻香村
跌水花溪
梯田景观
观演草坡
小桥流水
戏台廊桥

In design principles, it stresses the integrity and function of resort. The landscape design attempts to achieve the goals of nature ecology, detachment from the noise in the city and the ideal of leading a rural life. Meanwhile, the community function is introduced to satisfy people's communicating need. The design gives prominence to theme culture. The design has the characters of advanced quality, unique style, comfort, function, economization and standardization. What's more, the design puts people foremost and sticks to the principle of "people oriented" to show loving care for residents. The design also crates diversified living resort space

■ 亲子农场——播种收获的快意人生
Parent-child Farm —— A Pleasant Life with Pains and Gains

and Seeks for multiple levels of space. It also pays attention to ecology. It is in coordination with natural terrain and emphasizes the harmony between people and nature, architecture and nature and artificial environment and ecological nature. It makes full use of the terrain to create living environment and comfortable, leisurely resort space. It also has the characters of economization and aesthetics with reasonable landscape creating by using the natural and human conditions of terrain, landscape, water and plants, etc. the design puts emphasis on the combination of landscape effect and controlling cost and also on the region of plants' creating.

The combination with interest is stressed in functional activities and fun conception of farms and fields is added to the overall design. The plants design is also considered on a whole. More landscape elements are added on the basis of obeying the principle of integration to achieve a landscape effect of advocating nature appeal and delight.

■ 月光营地——相伴漫天繁星　Moonlight Camp — a pleasant life accompanied by the starry sky

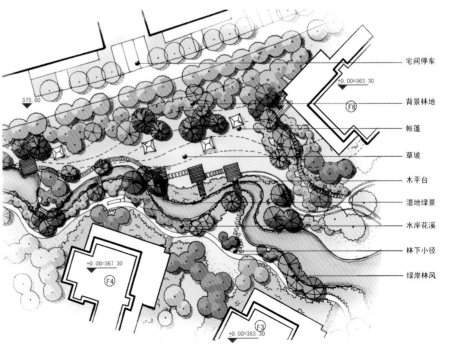

宅间停车
背景林地
帐篷
草坡
木平台
湿地绿景
水岸花溪
林下小径
绿岸林风

■ 月光营地放大平面　Enlarged Plan of Moonlight Camp

高层景观设计
High-rise Landscape Design

不断创新的主流设计观

长期以来由于我国人口密集的现状，高层住宅一直是地产发展的重要部分，并在今后很长的一段时期内还将继续成为大众选择的主流形态。尤其是最近几年，高层地产逐渐走向品质化、精致化，越来越强调人性设计、创新设计。业主对高层低密度环境以及恢复亲切邻里关系的向往也成为了设计师新的设计方向。绿茵景园在过去十几年中，通过与全国各地大型开发商的不断合作，积累了相当丰富和成熟的市场经验。本次我们选取了各种独具特色的案例展示并与大家共同分享。

高层类景观设计
High-rise Landscape Design

页码	项目
116	同盛·南桥 South Bridge By Tongsheng
124	天立·水晶城 Crystal City By Tianli
130	融汇·二期 Ronghui, 2stage
134	金科·黄金海岸 Gold Coast By Jinke
136	倍特·领尚 Brilliant Lead Noble Life
140	华润·橡树湾 Oky Bay Huarun
144	龙湖·三千里 Three Thousand Miles By Longhu
148	保利·云山国际 Cloud Mountain International By Poly
154	保利·公园198·丁香郡 Poly, 198 Park, clove Shire

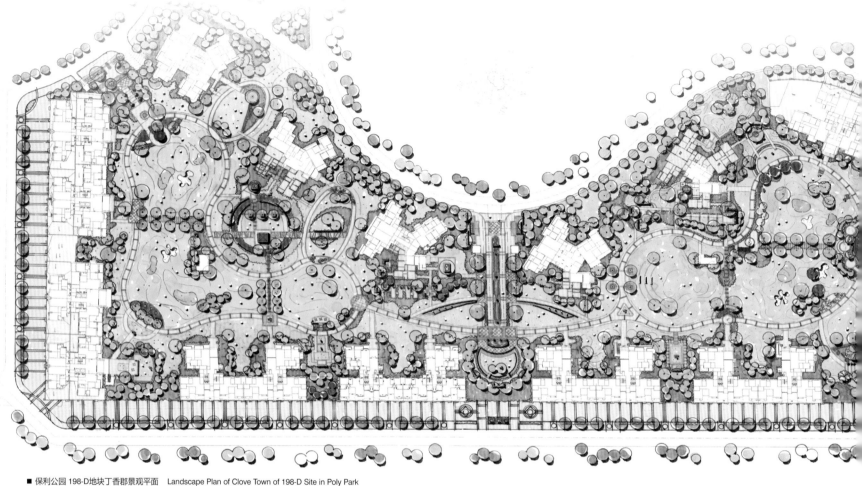

■ 保利公园198-D地块丁香郡景观平面　Landscape Plan of Clove Town of 198-D Site in Poly Park

宽庭私院的生活境界
富于现代自然气息的高层住宅

保利·公园198·丁香郡

The Modern High-rise
Residence with Capacious and Private Yard

Poly · 198Park · Clove Shire

项目开发商：保利(成都)发展有限公司	方案主创：阿笠
项目地址：成都北三环外蜀龙大道西侧	参与设计师：刘鸠鸠、张洪源、张悝
景观面积：60000平米	设计时间：2009年
设计单位：成都绿茵景园　设计二所	完成阶段：施工图设计中

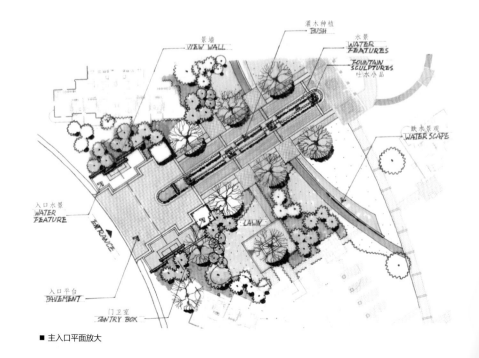

■ 主入口平面放大

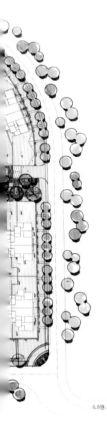

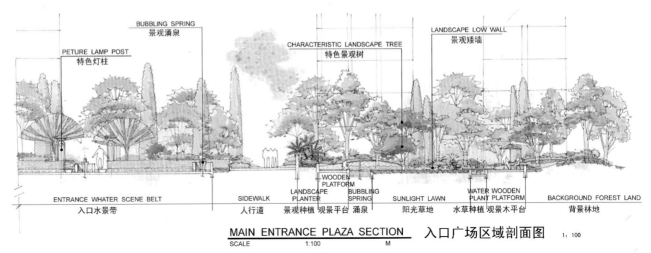

MAIN ENTRANCE PLAZA SECTION 入口广场区域剖面图
SCALE 1:100   1:100

- PETURE LAMP POST 特色灯柱
- BUBBLING SPRING 景观涌泉
- CHARACTERISTIC LANDSCAPE TREE 特色景观树
- LANDSCAPE LOW WALL 景观矮墙
- ENTRANCE WHATER SCENE BELT 入口水景带
- SIDEWALK 人行道
- LANDSCAPE PLANTER 景观种植
- WOODEN PLATFORM 观景平台
- BUBBLING SPRING 涌泉
- SUNLIGHT LAWN 阳光草地
- WATER WOODEN PLANT PLATFORM 水草种植观景木平台
- BACKGROUND FOREST LAND 背景林地

■ 主入口景观大道剖面 Main Entrance Plan and Enlarged Profile of Main Entrance Landscape Avenue

设计主题 Design Theme

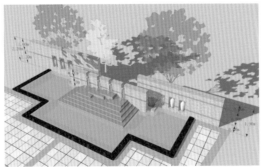

■ 主入口跌水池设计效果 Design Effect of Main Entrance Drop Water Pool

基地(丁香郡)属于保利公园198项目二期中的D地块，占地6万平方米。地块位于整个保利公园198项目较佳位置，左邻运动公园，右观30米道路绿化保护带及高尔夫景观，拥有较佳景观视线及宽阔视野，形成"外环境大视野，内景观长视距"。小区内未规划地面停车位，减少了日常交通对住户生活的影响，增加了景观绿化面积。架空层的设置有利于整合庭院景观，并提高休闲活动空间档次。景观空间的规划上，希望内部大庭院空间通过疏林草地体现；楼间围合空间通过密闭植物与休闲场地，增加宜人尺度的停留空间；归家空间上，通过植物的收放关系，以及重点植物的点缀，达到归家空间的亲切感。

■ 主入口空间设计效果 Design Effect of Main Entrance Space

■ 北面次入口台地效果　Platform Effect of Secondary Entrance in the North

Design Theme

The base (Clove Shire) is located in the D land plot, 2 stage of the Baoli Park 198 project, taking 60 thousand square meters. The land plot is at the comparatively superior location of the project, with sports park on the left and 30 meters greenbelt with golf court landscape on the right. It enjoys broad outer sight and far inner sig-ht. There is no parking lot in the residential area, thus it eases the interference on daily life of the residents and greening area is increased. The design of open floor is useful and upgrades the activity and recreational space. The planning of landscape space focuses scattered trees and lawns to give prominence to wide room of the yard. The surrounded space among buildings adopts plants and recreational area to create comfortable place for residents to stay. The space of residents way home stands out for its plants and amiability.

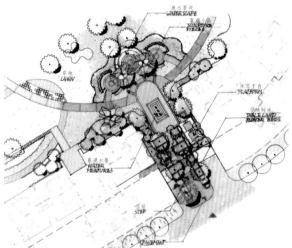

■ 北面次入口效果　Effect of Secondary Entrance in the North

■ 北面社区内景观广场　Landscape Square in the North Community

■ 北面林荫大道效果　Boulevard Effect in the North

■ 北面次入口水景效果　Water Landscape Effect in the Secondary Entrance in the North

■ 南面次入口效果　Effect of Secondary Entrance in the South

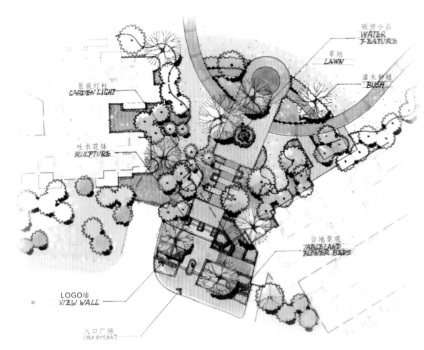

■ 南面次入口平面　Plan of Secondary Entrance in the South

■ 南面次入口立面效果　Elevation Effect of Secondary Entrance in the South

高层类景观设计 HIGH-RISE LANDSCAPE DESIGN

保利·公园198·丁香郡

■ 社区居民户外休闲空间 Outdoor Recreational Space for Community Residents

主题定位 Theme Orientation

公园中宽庭私院　　浪漫+轻松+纯净

景观营造策略：

空间策略：居在公园，院在居中

中央林丘公园带：道路、景观果岭草地、背景林地

架空层：泛会所式，居所功能的向外渗透，定位低于室内大厅

户外宅院：应用围合，提供简单使用功能，形成公共小花园

植物策略：阳光中庭，密语花园

中央景观果岭林地公园：造型草地、点景乔木、林地

户外交流空间：浓密乔灌木，提供静谧环境

有色植被：减弱草地单调，营造季节变化，丰富视觉层次

硬景策略：

视觉展示类：简练气质，整洁且有视觉层次

架空层：室内装饰情趣，居家品质的公共空间，精雕可品

户外交流空间：适用宜居的布局，舒适随意

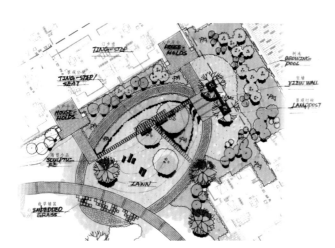

■ 亲情化的步径系统 Considerate Road System

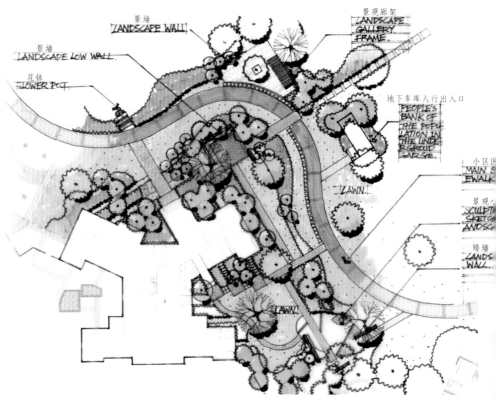

■ 节点空间放大平面 Enlarged Plan of Joint Space

■ 栋间空间设计效果　Design Effect of Space Among the Buildings

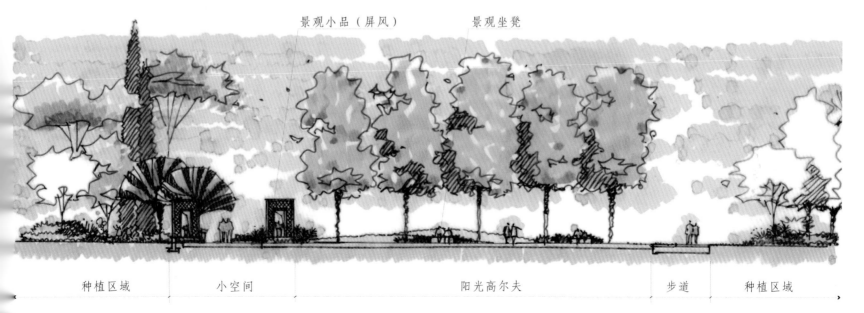

景观小品（屏风）　景观坐凳

| 种植区域 | 小空间 | 阳光高尔夫 | 步道 | 种植区域 |

■ 栋间空间立面效果　Elevation Effect of Space Among the Buildings

高层类景观设计 | 保利·公园198·丁香郡
HIGH-RISE LANDSCAPE DESIGN

■ 架空层景观效果　Landscape Effect of Stilt Floor

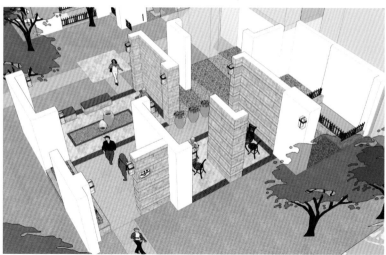

■ 架空层景观效果　Landscape Effect of Stilt Floor

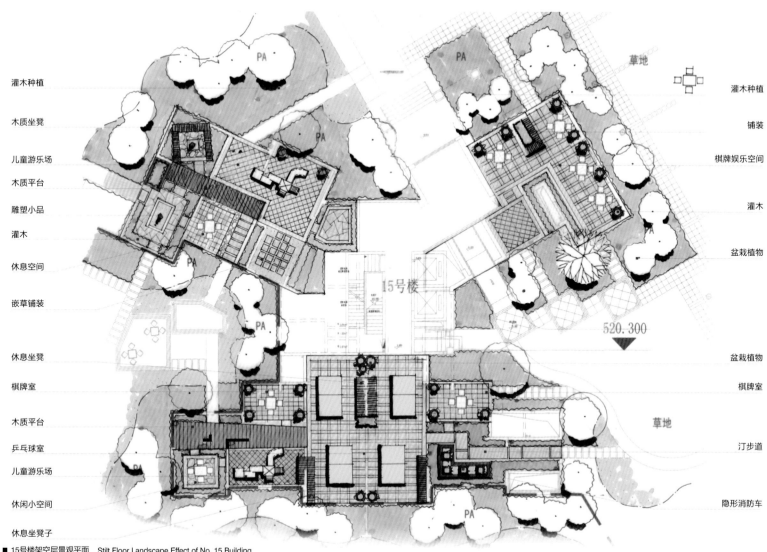

■ 15号楼架空层景观平面　Stilt Floor Landscape Effect of No. 15 Building

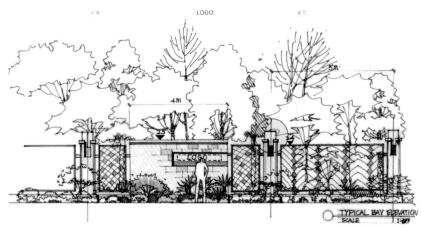
■ 社区标识围墙设计效果 Design Effect of Sign Walls of the Community

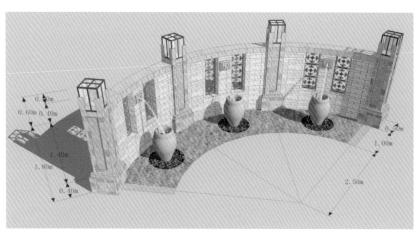
■ 社区内情景水景墙设计效果 Scenic Water Landscape Wall Design Effect in the Community

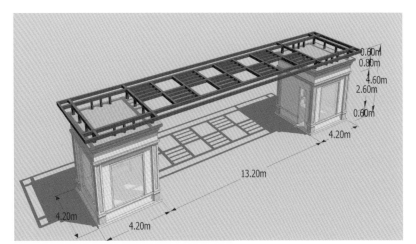
■ 社区主大门设计效果 Design Effect of Community Main Gate

The Capacious and Private Yard in the Park
Romantic Restful Pure

Landscape Design Strategies

Space strategy: living in the park with yard inside

The central wood park belt: roads + landscape fruit trees and lawns + forest background

Open floor: extensive chamber style, the extension of resident function and aims lower than chamber

Outer yard: application of enclosure, to provide simple functions and form little public gardens

Plant strategy: the mid yard with sunshine and little private gardens

The central landscape fruits trees park: the sculpted lawns and lightspot plants or woods

The outer space for communication: dense woods to provide quietness

Plants with various colors: to avoid the monotone of green plants and create seasonal diversity and enrich the sight

Hard landscape strategy: Outdoor living

Visual sight: concise, tidy and with different layers

Open floor: exquisite decoration and comfortable public space Outer space for communication: pleasant and comfortable layout

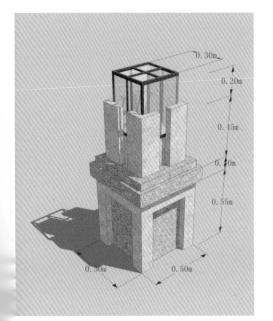
■ 标准化灯柱 Standardized Lamp Pole

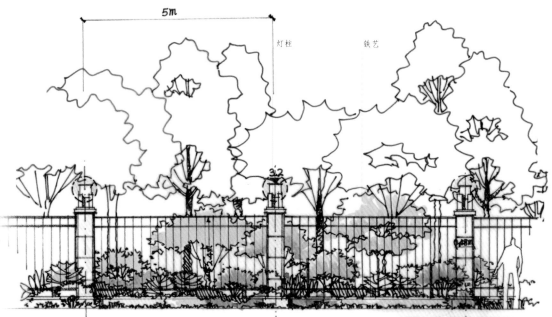
■ 社区围墙设计图 Design Effect of Community Walls

云中之旅
保利·云山国际

Journey in the Clouds
Cloud Mountain International by Poly

项目名称：贵阳保利云山国际居住小区
项目类型：现代风格居住区
项目面积：规划用地17万平方米
景观面积：15万平方米
设计单位：重庆蓝调国际（绿茵景园集团公司）设计一所
项目地点：贵阳市小关区
委托单位：贵阳海明保利房地产开发有限公司
主　创：任刚、陈云川、赖斯
团　队：张俊、李海平、毕领枫、林凤君、张予邯

项目简介 Project Introduction

在本案的景观设计细节中，我们运用了大量先进的技术措施来表达"云中之旅"的景观概念

① 时尚自然的社区内部环境 Fashionable and Natural Community Inner Environment

1、社区内部因车库及地形的不同标高，将社区分成了3个大的台地，在设计上我们运用了一种模块化的设计手法，将人们在外部空间的各种活动需求分成了若干的功能模块，通过重组构建的方式按照行为心理的需求，布置到景观大道的主轴中。由主轴串联起人们需求的主要活动模块，由下而上，由动到静设计出主要活动流线。

2、主轴活动的相对集中为我们在社区内保留了大量的绿色空间，这些绿地由周边的山林自然延伸到社区内，包裹住我们的高层建筑。建筑被绿色植物包裹的氛围，极大的增加了项目的生态价值。同时，大量的软质景观既保证了景观的高品质，又有效的控制建设成本。

In the designing details, we use a mass of advanced technical measures to express the landscape concept of "Journey in the clouds"

Fashionable and natural community inner environment

1. The community is divided into 3 huge terraces by it garages and landform. We adopt a measure of modularization and assume there are many modules for people's outdoors activities. There are rearranged and constructed in the landscape main axis according to the activity and mental demands. The main axis connects residents' major activity modules and form a flown line from below to the above.

2. The concentration of the main axis activities reserves for the community much green room, which naturally extends from the surrounding forest and encloses the high-rise building. It gives the impression that the construction are the plants grow in the forest and promotes the eco-value of the project. Meanwhile, masses of soft set ensures the high quality of the landscape and effectively decrease the costs.

② 野趣动感山地公园 The Wild Pleasure Dynamic Mountain Park

1、外围的高大岩壁将项目三面围合，原本是项目的最大障碍，我们在景观规划中希望赋予它新的内涵——海百合山地公园。（以贵阳盛产海百合化石而命名）

2、山地公园结合现状地形的自然岩体，将一条奇妙的探险之路暗布其中。景观设计手法将结合具体的景观节点，对其进行装饰。在难以处理接触的较高的岩壁，通过图案装饰等手法将其美化。在山体公园高处有一段可以坐看小关湖和黔灵山山顶的平台，在这里也可以鸟瞰整个贵阳城，不失为最能展现社区环境场所精神之地。

1. The surrounding cliffs enclose the project in three sides and was the biggest headache for the project, until we endows the project with new connotation: the Sea Lily mountain park, which derives from the abundant Sea Lily fossils in Guiyang.

2. A fantastic adventurous path goes through the mountain park and is among its natural rocks. Some scenic spots are decorated, e.g., on some rather tall cliffs, there are pictures. At the high part of the park, the Xiaoguanhu lake and the top of Qianlingshan mountain are visible. It is a key part of the park also for you can get a bird view of the Guiyang city.

③ 生态技术 Eco Techniques

周边山体岩壁及区内场地在雨季将汇聚大量的地表径流，景观在设计中将其收集集中，通过人工湿地等技术措施，将其净化成为社区内的景观用水。水净化系统，将弥补项目中对水景需求而带来的水资源不足，同时水景观系统本身也可以形成独特的景观风貌。合理的利用自然资源形成可持续化的生态景观，进一步的提升了云山国际先进人居理念。

当我们不再将景观单单作为楼盘宣传的噱头，而是实在还原它的作用和功能的时候，我们可以把看似纷繁复杂的景观解构成最原始的含义，正如千变万化的魔方一样，它其实是由一些最基本元素构成的。这些基本的元素支撑着我们丰富的户外活动和需求，外部的环境只是满足需求的空间载体。景观的构思来源植物的叶脉，由一条主轴串联起若干的功能模块。

我们试图重构我们生活中所需求的最原始的景观诉求，通过现代的材质和构成方式来重构一座离尘的新城。

A great deal of earth surface water is collected in rain season and turns to landscape water through artificial everglade and purification. The purification system is a unique view and compensate for the shortage of water in landscape. To use the natural resources and form sustainable landscape upgrades the project's definition of harmony.

When we no longer take the landscape as the stunt of real estate advertisement, but to revert its original functions, we may, see the seemingly complicated definition as its original meaning. Just as the magic cube is actually consist of basical elements, which meet our colorful outdoor activity demand. The design inspiration originates from the vein of plants and a main axis links all the functional modules together.

We are making effort to reconsider the basical requirement for landscape and also rebuild a new city of arcadia by modern materials and measures.

高层类景观设计 | 龙湖·三千里

HIGH-RISE LANDSCAPE DESIGN

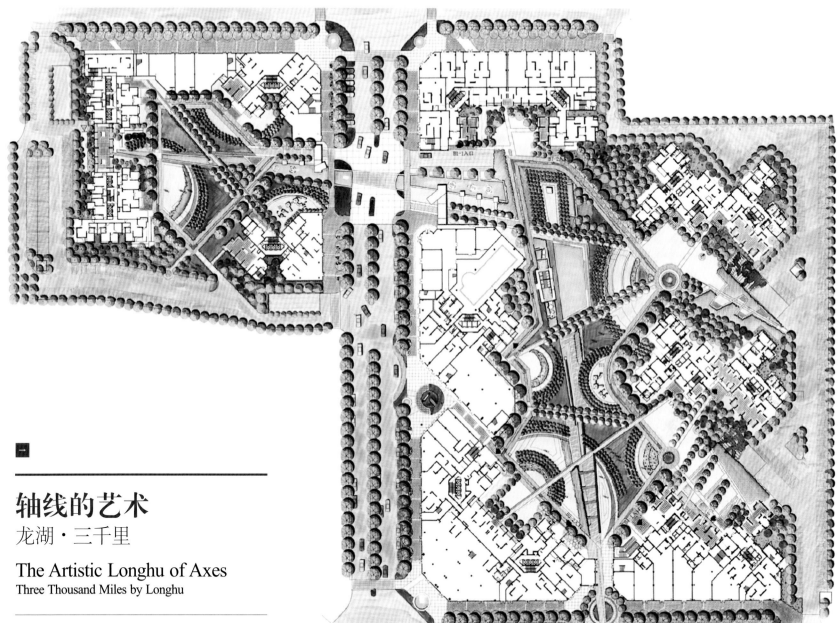

轴线的艺术
龙湖·三千里

The Artistic Longhu of Axes
Three Thousand Miles by Longhu

开发商：成都龙湖锦华置业有限公司
地　点：成都城东成华区二环路东二段5号
规划用地面积：5.5万平方米
景观面积：4.2万平方米
项目类型：现代简约风格居住区
设计单位：成都绿茵景园 设计一所
主　创：潘旭、曾繁利
团　队：粟凡粒、张彬、肖浩波、陈浩然
设计时间：2008年

设计主题 Design Theme

"三千里"项目景观设计旨在与建筑"现代简约"的整体设计概念相匹配。创造具有视觉冲击力的景观效果,给人以深刻印象。整体色彩强调清新自然、明快雅致,利用色彩及材质的对比形成视觉上的亮点。体现出现代高层住宅简约、时尚的感受。该项目已发展成为片区具有优秀居住环境、创新的居住空间和标志性外观的居住社区,实现项目总体市场效益、目标经济效益、社会公众利益和城市环境品质的多种价值取向的统一。

项目的景观设计理念就是要营造一个现代的、生动的及动态的景观,以充分反映目标购买者前卫的特性。不同于许多竞争项目中一般采用的由软景围绕大型湖泊的做法,此项目的设计理念就是要在布局中采用主导元素,包括水体、小径、软景、墙体,重新演绎这些元素之间的关系,并通过这种做法营造一种现代景观。这些元素通过水轴、墙面、小径线路及不等边四边形的软景来表达。元素之间相互重叠、交叉、替换,形成了一系列相互联系的动态空间。景观设计就是要将分别位于市政道路两侧的两个地块连成一体,在从33层塔楼向下观望时可解读成一个抽象的三维艺术形式,并为在地面上行走的人提供一系列丰富的体验。景观通过运用现代的组织结构将两个地块连接起来,并形成了两条主要的组织轴线来表达设计理念。这两个轴一个为东西向,另一个为南北向。东西向的轴线由将市政道路两侧的1号及2号地块连接起来的水轴位面控制。这个轴突出了至每个地块的主要入口,并将主要的入口大门设置成与水体相融的建筑小亭。南北向的轴在东部地块大型的庭院空间中延伸,将次入口与该轴线的东西向轴线交叉点相联。南北向轴线的特点为分段的水轴及断断续续的线形墙体元素。外部的游泳池、消防水库及水池设施大楼与这个带状的轴相结合,形成了一个设计特色,并强化了轴线的组织结构。一系列线形的步行小径与这两个主轴相交,并定义大小不一的弧状景观区,连接一系列空间,而这些空间包容了各种景观特色区、活动功能及景观主题。这些景墙通过与其他主要的景观元素相交。在这些交叉处,墙体将形成一个偏移轴,展现了设计中的空间被划分的特性。轴向的几何形设计延伸至周边的商业街,已在两个主要庭院的内部空间与项目公共外围之间形成一种关系。

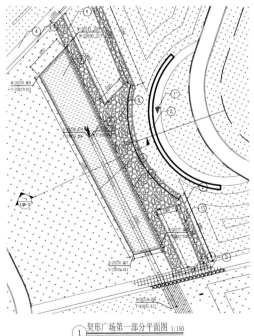

① 100x100x20mm厚灌县黑烧面
② 木座凳 详见LD1.02-3
③ Φ30~50mm雨花石、黑色、抛光面
④ 400x400x20厚黄木纹板岩拉裂面
⑤ 20厚锈石黄花岗石碎拼(D=200~400缝宽15~20mm)
⑥ 龙墙详见 LD1.08-2
⑦ 600x600x20中国黑花岗石光面
⑧ 300x300x20中国黑花岗石光面
⑨ 300X210X20中国黑花岗石光面压顶

① 契形广场第一部分平面图 1:150

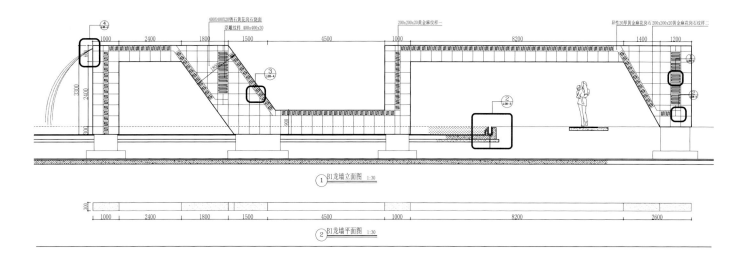

① B1龙墙立面图 1:30

② B1龙墙平面图 1:30

■ 几何形的景观雕塑与平面几何的构成相呼应，花池的花径又极好的柔化了这一种几何体的生硬。
The geometric landscape sculpture combines with plane geometry and the road with flowers of the colored pond softens well the rigidity of the geometric combination.

Design Theme

The design intention of "Three thousand miles" is to go in accordance with the modern and concise style of the architecture. The design is natural, bright, and good at forming materials contrast. The project aims to be an outstanding and innovative living community and achieve the goal of fulfilling market, social and environmental values.

The design conception is to create modern, lively and dynamic landscape and reflect the customers are avant-garde. Compared with the rivals' designs of enclosing the lake with soft set, it intends to create modern landscape with water, paths, soft set and walls. These elements are presented by water axis, wall surface, paths and soft set of unequal quadrilaterals.

They overlap, cross and replace each other and form a series of correlative dynamic space. The design connects the 2 land plots on the each side of the municipal road together by modern structure and looks like an abstract 3-dimention artistic form while viewing from the 33th floor. Meanwhile, it also provides pedestrians with colorful walk experience. There exit 2 main structure axes to interpret the design conception, east-west direction and south-north direction. The east-west axis is controlled by the water axis which connects the NO1 and NO 2 plots beside the municipal road. It gives prominence to the main entrances of each plot and by the water the harmonious pavilions serve as the main gates.

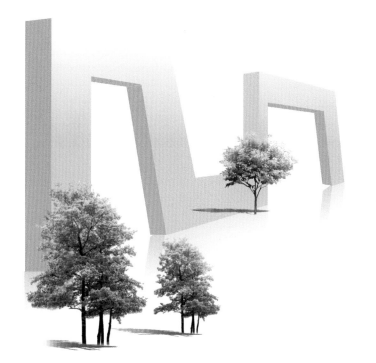

The south-north direction axis extends to the huge yard space of the east land plot and connects the secondary entrance and the crossing points of the east-west direction axis and itself. Its features are water axis and the snatchy line-shaped wall element. The outer swimming pool, fire control reservoir and the water facility building combine with the axis and strengthen its structure. A series of line-shaped paths cross the two axes and connect space of various sizes. At the crossings the landscape walls become an offset axis and display the space is divided.

设计概念的总体效果是要营造一种独特的现代景观，可展现该景观构成元素相互交叉形成的复杂性，同时为居民及访客提供一种难以忘怀的体验。

The overall effect of the design is to construct a unique modern landscape and present the complication of the consisting and crossing elements, meanwhile, it also aims to provide an unforgettable experience for residents and visitors.

英伦学院式园林情境
华润·橡树湾

English College Garden Style
Oak Bay by Huarun

项目开发商：华润置地成都发展有限公司
项目地点：成都市犀浦西区大道
设计面积：21500平米
设计单位：成都绿茵景园 设计二所
景观方案设计：韩茜
景观施工图设计：马卉
发展阶段：2008年景观设计
完成阶段：施工完成

■ 自然式外围绿化带 Natural Outlying Greening Belt

■ 自然式外围绿化带 Natural Outlying Greening Belt　　■ 尊贵感的社区入口 Noble Community Entrance

■ 主入口特色水景 Feature Water Landscape of Main Entrance

设计特色 Design Features

根据建筑所营造含蓄、典雅的人文气质和学院气息，打造具有独特设计品质和创新理念的外围景观带，以国际化、人性化、本土化的原则对英伦学院式园林情境进行全新诠释，创造强有力的视觉冲击力及亲身感官体验；形成品质社区的亮点，同时也成为良好的城市景观带。

In accordance with the elegant humanism and college atmosphere of the architecture, the surrounding landscape belt focuses on uniqueness and innovation. With the principle of being international, human and local, the design endows the English college garden style with new definition and creates a powerful visual impact. It is a lightspot as well as the excellent city landscape belt.

高层类景观设计 | 倍特·领尚
HIGH-RISE LANDSCAPE DESIGN

■ 变化丰富的泳池剖面图　Profile Drawing of Swimming Pool with Various Changes

LOHAS LIFE

流动的景观，带动着建筑翩舞……
乐活的景观，带动着心灵翩舞……

→

领舞新城，乐活风尚
倍特·领尚

Lead the Dance in the New City and Enjoy LOHAS way of Life
Brilliant Lead Noble Life

开发商：绵阳倍特房地产开发有限公司
地点：绵阳高新区
设计面积：48610平米
设计单位：成都绿茵景园 设计二所
主创设计：阿笠、韩茜
方案团队：张洪源、刘鸠鸠、邹巧都
施工图设计团队：马卉、段倩、袁亚宁
发展阶段：2009年景观设计
完成阶段：已施工完成样板区

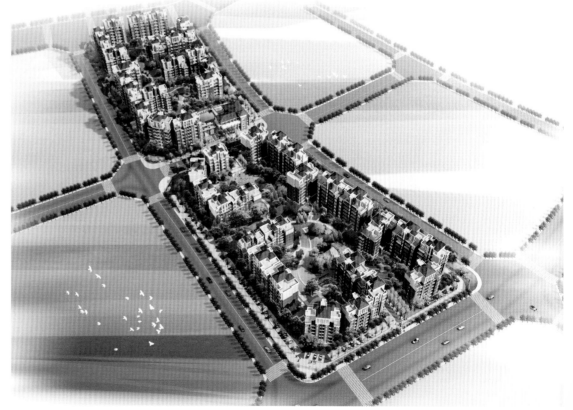

■ 鸟瞰图　Aerial View Drawing

→ 设计主题 Design Theme

倍特·领尚是位于绵阳市政府着力打造的"园艺之巅"第一高尚居住区，总建筑面积10万余平方米。毗邻5000亩原生态缓坡山林、120亩雷锋湖、西山风景区等众多原生态绿化区域，生态环境清幽；周边三所私立贵族学校、五大高等教育学府环绕，人文气氛浓郁；九洲大道和剑南路交汇于此，交通网络四通八达。

Lingshang is located in the first high-grade residential area "At the top of Gardening" which is a work of Mianyang government. It is over 100 thousand square meters and next to the 823 acres pristine forests on gentle slope, 20 acres Leifeng lake, West mountain scenic area and many other primitive greening area. There are 3 private schools and 5 colleges in the vicinity with atmosphere of humanity. Here is a crossing of Jiuzhou Avenue and Jiannan road and is easy to get everywhere.

设计主题：悠然闲逸，适意自得 —— 乐活家的生活哲学

核心竞争力：定制组团私享客厅，引领乐活格调生活方式

景观风格：流动的美学，精致回归质朴的浪漫，自在的优雅

景观语言：乐活、浪漫、优雅、健康、流动、惬意

Design theme: leisure, comfort and pleasure —— the life philosophy of "LOHAS"

Core competitiveness: making private living rooms of the divisions and leading the life style of "LOHAS".

Landscape style: flowing aesthetics, exquisite and simple romance, comfortable elegance.

Landscape description: LOHAS, romantic, elegant, healthy, flowing and comfortable.

■ 无边界泳池图　Plan of Boundless Swimming Pool

■ 儿童游泳池　Swimming Pool for Children

■ 园区曲折自然水景图　Winding Natural Water Landscape Drawing in the Garden

■ 入口门庭图　Drawing of Entrance Courtyard

高层类景观设计 | 倍特·领尚

■ 新古典风格会所门庭效果图　Effect Drawing of Club Courtyard in a New Classical Style

■ 围墙整体为铁艺与实体墙结合，通透的墙面在转角的区域，陡然转变成为精致的实体墙面，形成社区的地标
The wall integrates blacksmith and solid wall. The hollow-out wall suddenly becomes a solid wall in the corner, which forms the landmark of the community

设计中融入先进的乐活生活理念，LOHAS（乐活）是一个西方传来的新兴生活型态，意为以健康及自给自足的形态过生活，强调"健康、可持续的生活方式"。

"健康、快乐、环保、可持续"是乐活的核心理念。

"乐活"是一种环保理念，一种文化内涵，一种时代产物。它是一种贴近生活本源，自然、健康、精致的生活态度。从产品风格和生活方式整合定位，与绵阳第一居住区园艺新城的自然风貌、人文风情相得益彰，予以建筑一种文化和内涵，演绎一种浪漫温馨。舒缓闲情的生活情调和韵味，勾勒出稀缺的人居高地时尚山居生活方式，带给人很多美好的向往；让人依恋的不仅仅是房子，更深层次的是对生活环境、生活方式、生活韵味的憧憬和想象。

设计师巧妙运用流动的美学，景观构图如图画般展开，即使是静态，也如同流动一样具有张力，寓动于静，相得益彰。

The project is combining with the concept of LOHAS, which is a western newly advocated life style and means lifestyles of health and sustainability.

Health, happiness, environmental protection and sustainability is the core concepts of LOHAS.

LOHAS is an environmental protection conception, a culture and a product of the time. It is a natural, healthy, and delicate living attitude close to life. Oriented from the project and living style, and combining with the natural and historical scenes of the Mianyang residential area, the landscape design endows the architecture with a connotation of culture and showing a romantic and restful living style. This rare and fashionable upland way of living brings about beautiful expectation and love for living environment and living style.

The designer skillfully adopts the aesthetics of flowing, presenting the landscapes as pictures in steps. Even if the static scenery is with the attraction of flowing.

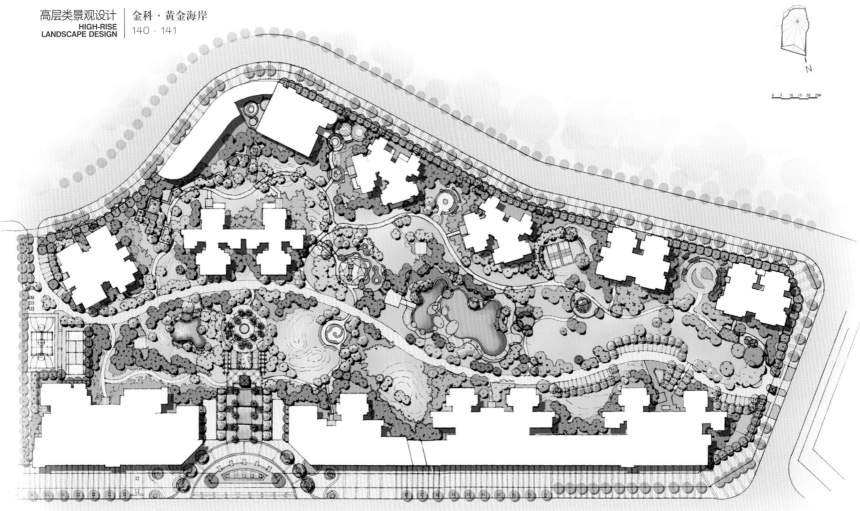

■ 金科廊桥水岸总平 General Plan of Jinke Gallery Bridge

绿色生态的阳光地中海风情住区

金科·黄金海岸

The Green and Eco Residential
area with Sunshine of Mediterranean

Gold Coast by Jinke

项目类型： 居住区
项目面积： 规划用地110亩
景观面积： 5万平方米
设计单位： 重庆蓝调国际（绿茵景园集团公司） 设计三所
项目地点： 重庆涪陵滨江大道
委托单位： 重庆金科集团
主　　创： 张勇
团　　队： 肖勇、李锦香、郭怿、彭盼、汤元英、徐秋林

项目简介 Project Introduction

项目位于重庆市涪陵区滨江路，背山临江，地理位置优越，景观资源丰富。

景观围绕四个鲜明的特点进行设计：

1、阳光草坪、疏林绿地为主的自然格局；
2、丰富完善的住区功能、驻足交流广场、邻里休息空间、漫跑步道、运动场地、老年儿童场地、风情泳池等功能充分的布局，满足住户丰富的日常生活；
3、以水为主题，在注重借景长江的同时，在主入口和住区绿地内，也将水作为设计的重要元素，引入业主的生活空间；
4、精致的异域风情细节和小品元素。

前期建设完成的样板展示区，景观面积2万平米。分为主入口景观轴线、中心广场、纯美草坪三个部分。

在与市政道路相连的主入口，以气势的广场水景吸引人们的视线，丰富城市景观。水，是主入口景观的主题，水的魅力在这里得到淋漓尽致的展现。开阔的弧形入口空间，布置涌动的水景和葱茏的榕树，向人们敞开迎宾的怀抱。跌水、喷泉环绕在绿色灌木和花带中，带给住户无限美景。中心广场是整个入口景观轴线的高潮。四面环绕的潺潺流水、弥漫花香和高大的老人葵，为住户创造美丽动人的交往空间。穿梭在蓝色的花丛和绿草树林中的蜿蜒小道，将人们带入了迷人的绿色丛林深处。一簇簇清爽的白桦，高大的水杉，让你抬头间感受蓝天与阳光。

后期建设中的老年儿童游乐场、风情泳池、宅间花园、漫步花园都将围绕这些景观而精心打造。带领居住者感受异域风情的浪漫情怀，创造适合现代人居住的美好生活环境。

■ 中心水景 Central Water Landscape

■ 主入口 Main Entrance

■ 跌水景观 Drop Water Landscape

■ 园路停车效果 Parking Effect of Garden Road

■ 门卫室 Guard Room

The project is located at Binjiang road, Peiling district, Chongqing. It has mountain background and beside the Changjiang river, enjoying the superior geographical location and abundant landscape resources. The design gives prominence to 4 points:

1. The natural layout of lawns with sunshine and scattered trees.

2. Multiple community functions, such as communication square for residents, restful space, jogging lanes, sports fields. Space for the seniors and children, swimming pool, etc., to meet the living demands.

3. Adopting water as a theme: besides the scene of Changjiang river, to use water as an essential element and introduce it to the living space.

4. Exquisite exotic flavor and elements.

■ 浅塘　Shallow Pond

The sample area landscape of the former stage takes 20 thousands square meters, dividing into entrance landscape axes, the central square and the fascinating lawn.

At the main entrance square connected with the city avenue, the magnificent scenery of water is definitely appealing. It is the theme of the landscape and its charm is presented to the full. The open arch entrance, the flowing water and the verdant banyans are hospitable to the residents. Little waterfalls and springs are surrounded in the bushes and flowers. The central square is the climax of the entrance landscape axes with surrounding brooks, floating flower fragrance and tall Washingtonia robusta, all of which create an ideal space for communication. The winding paths in the woods lead to the inviting deep in the jungle. Up ahead, are the gorgeous birches and lofty metasequoia with blue sky and sunshine.

The latter stage landscapes are still in construction. The playground for the seniors and children, swimming pool and gardens will be in accordance with the former one and bring about exotic flavor and create an amazing and livable place for modern people.

高层类景观设计
HIGH-RISE LANDSCAPE DESIGN | 融汇·二期

韵动的风景
融汇·二期

The Scene with Rhythm
Ronghui, 2 Stage

开发商： 融汇地产集团
项目地址： 重庆市巴南区李家沱融汇半岛
设计单位： 重庆蓝调国际（绿茵景园集团公司）设计二所
设计时间： 2009年
竣工时间： 2010年5月
投资金额： 450万
景观面积： 18000平方米
设计师： 王轶、林凤君、樊菁、潘启渝、刘盛义

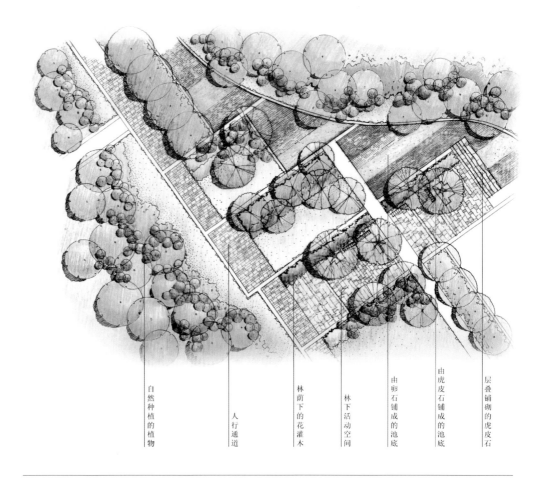

自然种植的植物 | 人行通道 | 林荫下的花灌木 | 林下活动空间 | 由卵石铺成的池底 | 由虎皮石铺成的池底 | 层叠铺砌的虎皮石

设计主题 Design Theme

高层建筑的围合让A组团的中庭形成了一处独立的社区空间。设计师思考此种集中式活动场所的营造方式应该是以区域中心作为视觉景观的焦点、公共、开敞，而社区内的活动都围绕中心景观来开展。

因此，我们把高层围合下的空间作为组团的中心绿地，中心大草坪像绿色的宝石一样成为组团内最清爽、最直接的部分。草坪西侧靠近建筑的区域是组团集中的活动场所。分枝较高的合欢树作为场所顶界面，创造舒适、宜人的林下空间。

人们在视线通透的场景下自由地展开活动：聊天、休憩、散步、谈恋爱……活动场地与中心草地相连接的部分形成了一条带状区域，我们采用水体作为连接场地与中心草地之间的媒介。柔美的线条很好地过渡了林荫下硬景的块面感。蜿蜒韵动的水体成为中心一条唯美的"丝带"。它正好同建筑刚硬的外壳对比强烈，即使是在水体最柔弱的部分，其形体也能表达出环境的张力。同时，水体被设计成两种效果：枯溪和水溪。既考虑了景观价值，又很好地结合了管理层面的操作。而另一侧，展现在片状空间里树影斑驳的草坪和树林，营造了草坡上自然、舒朗的绿地环境。漫步林中，鸟语花香，趣意幽然。

■ 块面感强烈的平面布局为草坪和树林所柔化
Plane layout with a strong sense of cubes and planes is softened by grassland and trees.

Design Theme

The high-rise buildings of A group surround an independent community space. The architecture designer's intention for the centralized activity site is to take the space center as the sight focus. All the community activities will be done around the central landscape.

Thus, we take the space as the central green area of A group. The central lawn resembling a jade as the clean, and chief part of landscape. The west part of lawn, in vicinity to the buildings, is the main area for activities. Albizzia with high branches are used as the top of the space and create comfortable and pleasant shadow room.

Chatting, resting, stroll, dating etc. can be found in the bright space. A belt connects the activity area to the central lawn and we take water as the medium. The soft lines of water serve as an excellent transition for the hard landscape. The water is in the shape of a gorgeous silk belt, winding and rhythmic, in sharp contrast with the rigid and hard buildings. The water is designed in several effects, such as brooks, so that it is with landscape value and easy for management. On the other side, the green lawn and the trees give a perfect place to enjoy the bird twittering and plant fragrance.

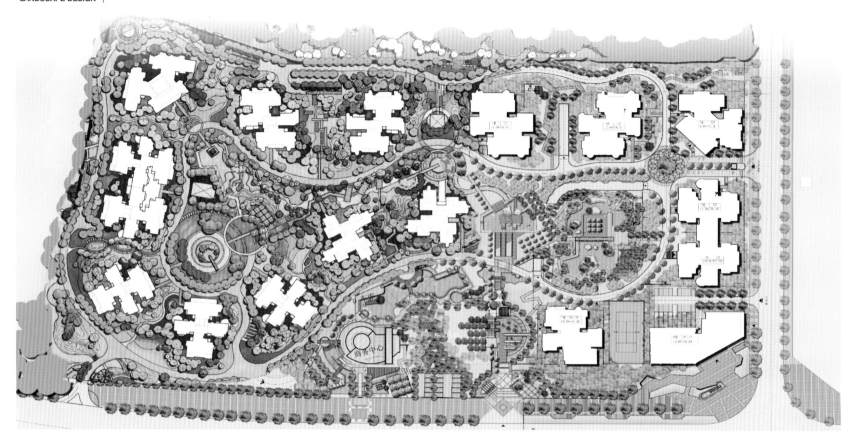

设计感受生活
天立·水晶城

To Feel Life in the Design
Crystal City By Tianli

项目开发商：四川天立有限公司
项目地点：四川泸州江阳西路景观大道
项目面积：规划用地150亩
景观面积：5万平方米
项目类型：现代风格居住区
设计单位：重庆蓝调国际（绿茵景园集团公司） 设计一所
主　创：任刚
团　队：陈云川、毕领枫、赖斯、李海平、张俊

主题定位 The Theme Orientation

天立·水晶城，泸州市城西新区景观大道上的地标建筑群落。位于市政规划建设的"集旅游、生态、居住为一体的生态休闲住区"，政府倾力打造的公园样板区斑竹林水库近在咫尺，毗邻近千亩城西市政生态公园。

当我们不再将景观单单作为楼盘宣传的噱头，而是实在还原它的功能和作用的时候，我们可以把看似缤纷复杂的景观解构成最初的含义。设计中我们引入魔方的概念，纷繁的景观元素被有机重构成不同的景观现象，形成令人印象深刻的有趣空间。

通过"魔方"概念，水晶城的环境被塑造成独一无二的城市绿洲，这个绿色魔方有着严格组合标准：材质上，有机无机相结合；质地构造上，从粗糙到精致；明暗配合上，介于明与暗之间。照明设施，简洁的灯具与柔和的光源形成富有诗意场所，种植方面，因地就势的地形与一体化的植栽将小区的各个层面结合在一起，给游历小区的游客带来独特的瞬间。这些风景特色，只有当你切实在其中穿行时才能感受到。从上方可以看到花园的整个布局。

迎合项目的特质，选择了一些通透明快的材料，如玻璃等被大量运用。

值得一提的是这些新颖的景观，从某种意义上给城市和地产带来了一些活跃的东西，改变了人们对景观的理解。同时，对项目销售和口碑也产生了意想不到的效果。

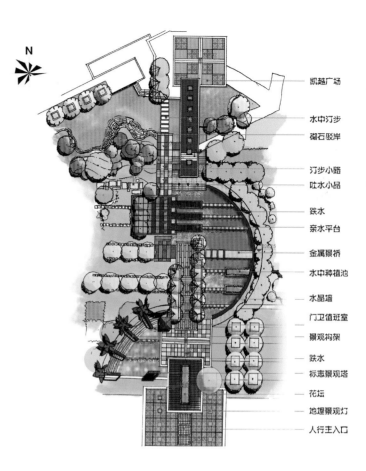

凯越广场
水中汀步
砌石驳岸
汀步小路
吐水小品
跌水
亲水平台
金属景桥
水中种植池
水晶墙
门卫值班室
景观构架
跌水
标志景观塔
花坛
地理景观灯
人行主入口

The Theme Orientation

The Crystal City by Tianli, the landmark construction in the west new district of Luzhou, is located in the multi-functional eco-residential area for travelling, ecological, and living according to city planning. The sample park Banzhulin Reservoir planned by the government and an enormous eco-park are in the vicinity.

At the time we cease to take landscape as the stunt of real estates advertisements, and vivify its practical functions, we comprehend the seemingly complicated landscape as its original meaning. In the design we introduce the concept of magic cube, in which the various landscape elements are recomposed into different sceneries and form an impressive space.

By the concept, the environment of the crystal city is fashioned into a unique urban oasis. The green magic cube is with strict standard for assembly: in the material, both organic and inorganic are adopted; in textures, both rough and delicate are employed; in the color and light, it is a product between dark and bright; in the lighting facilities, concise lamps and soft light form an romantic place; in plantation, it combines the original landform and integrative plants. The feeling for these landscape characteristic are so unique and can only be felt when you are actually in it. From above, the layout of the garden can be seen.

In accordance with the above features, the materials for landscape are mainly transparent, and glass is widely employed to from peculiar sceneries.

What is worth mentioning is that these innovative landscape brings about vitality to the city and real estates, changes people's views of landscape. It is significantly helpful to the sales and brand establishment.

■ 别致而新颖的水体设计让人眼前一亮
Exquisite and fresh water design impresses people

高层类景观设计 | HIGH-RISE LANDSCAPE DESIGN
天立·水晶城

■ 玻璃的大量运用使景观界面断而不闷、通透明快
Abundant use of grass makes the landscape boundary penetrating but not stuffy, clear and lively

■ 在大体量丰富运用的同时，细节的打造也明快精致
With the abundant and rich use of large bulk, the creating of details is also lively and exquisite

邂逅简约，品味奢华
同盛·南桥

Encountering Conciseness and Tasting Luxury

South Bridge by Tongsheng

开发商：同盛集团公司
项目地址：上海市南桥镇航南路陈桥路
项目规模面积：28000平方
项目类型：大型社区住宅楼盘
景观设计：上海绿茵景园
主创团队：余志国、刘美丰
设计时间：2009年十月至今

■ 商业街一角　A Corner of the Business Street

简洁干练的欧式元素迸发出简欧风格的缩影

"硬朗、大气、内敛、高雅、尊贵、简约的奢华………"

→ 设计阐述 Design Introduction

现代景观呈现出多元化趋势的同时,让我们看到简洁、大气、清新、的景观仍就屹立挺拔。有人说这是一道独特的风景,是新时代简约主义的变革;有人说,这是对生活态度的新诠释。"邂逅简约,品味奢华",以简欧风格作为构架,营造出简约而不简单、奢华而不奢侈、淡定自若、大气硬朗、尊贵典雅的景观规划。

In contemporary society, when landscape designing unfolds diversity, the concise and splendid landscape still stands out. Some people contend it is a unique view as well as a reform of minimalism of a new epoch, while some hold it is a new interpretation of living attitude. "Encountering Conciseness and Tasting Luxury", with this goal, the project takes the terse European style as a frame and create a dignified and tasteful design.

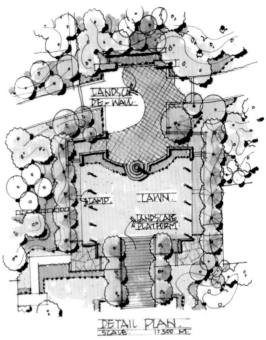

■ 入口详图　Detail Drawing of Entrance

■ 风格现代大气的主入口　Main Entrance in a Modern and Splendid Style

高层类景观设计 HIGH-RISE LANDSCAPE DESIGN

同盛·南桥

■ 河滨通道详图　Detail Drawing of River Channel

其设计概念及思路沿用建筑的设计手法，讲究环境设计与建筑的紧密配合，并增加了许多欧式豪华的元素，简洁不失豪华，其效果强烈，给人强大的冲击力。环境设计上接近自然，有值得品味和揣摩的风格，实用但不造作，看上去细腻而又经得起时间的考验。其设计特点是：体现欧洲精致生活情调，融建筑和居住为一体的整合景观，无限接近自然、以人为本，体现人文关怀。设计师通过社区建筑、入口、街景、景观、小品以及近人的尺度感倾心营造了一个温馨、和谐、典雅、富于趣味性的现代欧式风情小镇。丰富多样的四季植物景观和灵动的点状水景以及休闲的户外空间是小镇景观的特色，空间规划清楚并极有层次感，从景观视线到空间、

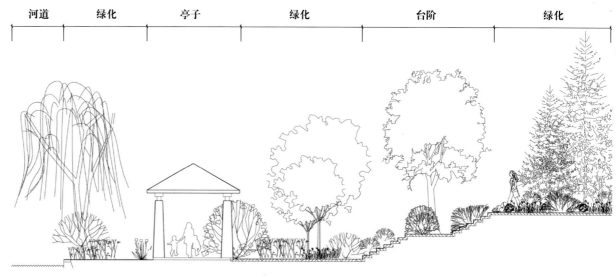

■ 河滨通道剖面图　Profile Drawing of River Channel

休憩花园、游乐场地都合理的安排于小区内，意在为居住者创造一种明快、丰富而多层次的户外空间结构，提供一种健康休闲的居住氛围，使人感觉到生活的情趣。景观从室外延续到室内，与建筑现代、简约的风格相呼应——不仅仅是户外色彩丰富、形式多样的园林，也不仅仅是颇具小资情调的豪华家园，大自然的元素被引进起居空间并成为建筑和生活的一部分。

The landscape design is in accordance with the architecture and focuses on the combination, with a considerable of European elements which are luxurious and concise and form a sight impact. It is natural and practical and enduring in time. The characteristics are to present the European life style and combine the architecture with landscape and to be close to nature and resident-oriented. The designer creates a comfortable, harmonious and interesting modern European town with community buildings, entrance, streets, landscape and parergon. The diverse plants of four seasons, water scenery, and the restful outdoor space are the characteristics of the landscape. Its space planning is clear and with many layers. Landscape view lines, space, gardens and playground are all included in the residential area and it intends to create a colorful and multiplelayer outdoor space and healthy living atmosphere. In line with the modern and concise architecture style, the landscape is colorful and introduces the natural elements.

■ 儿童活动空间效果图　Effect Drawing of Activity Space for Children

■ 花园组图空间效果图 Effect Drawing of Garden Space

■ 交通节点效果图 Effect Drawing of Traffic Joints

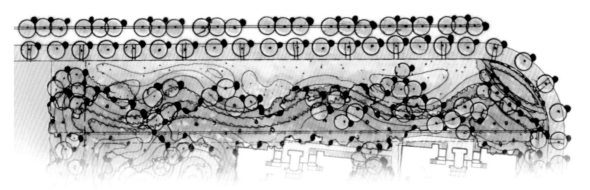

DETAIL PLAN.
SCALE: 1:300 M.

　　具体设计上，保证园区主要消防道路的畅通。在庭院和入户道路的处理上，以满足人行为标准。消防登高面的处理，以保证消防要求为前提，运用道路基础与草坪的处理方式，尽可能的减少硬质道路的面积，保证整个小区绿化面积最大化。

　　In detailed design, the fire control lane is assured of being unimpeded. In the design for yard and the entrance road, resident's convenience is given prominence. After meeting the requirement of fire control, the adoption of lawns maximizes the green area.

■ 市政绿化带详图 Detail Drawing of Municipal Greening Belt

综合性景观设计
Synthesized Landscape Design

多样性整合的景观典范

综合性景观由于其自身功能的多样性，对园林设计的要求显得更为苛刻：既要满足多种功能，兼顾优劣势，又要营造出赏心悦目的景观氛围。大型综合性景观还要求契合城市整体规划，立足长远发展战略，协调该项目各种规范要求。综合性的景观设计还特别强调设计风格的整体性和完整性，对与景观相关的各个领域都要有所涉猎，无论对公司还是团队都有着更高的要求。综合性景观将成为设计师们攀登的新的高峰。

综合性景观设计
Synthesized Landscape Design

| 162 | 166 | 174 |

仁恒置地广场
Renheng Zidi Square

富临·桃花岛
Fulin.peach Flower Island

天津泰达·上青城
The Front Qingcheng By Taida,Tianjin

■ 与建筑立面呼应的售楼部前场广场
The Square in front of Sale Department in Combination with the Architecture Elevation

山的印象
天津泰达·上青城

The Mountain Impression
The Front Qingcheng by Taida, Tianjin

项目名：青城山泰达上青城（形象样板区）
地址：都江堰市青城山前山大门1500米处（成青快铁终点站）
面积：占地526亩
设计单位：成都绿茵景园 设计二所
设计时间：2010年-至今
主创设计师：韩茜
参与方案设计师：邹巧都、潘旭
硬景施工图团队：曾凡利、姚舒雅、阙龙庆

■ 嵌入式浪漫花田休闲区　Embedding Romantic Recreation Area of Flower Field

景观主题 The Landscape Theme

山的印象——幽谷坡地公园 Glen Park Slope 景观设计灵感——源于建筑的肌理和造型。意在通过建筑立面与景观平面的呼应衔接，采用几何形造山理水的方式，创作出趣味与意境相结合的景观空间。

景观设计特点——大地艺术、无界景观

■ 场景图　Scenic Drawing

■ 形象样板区总平面图
General Plan of Image Sample District

The Landscape theme

The mountain impression: the valley and slope park.

The design inspiration: it originates from the architecture style and structure. Through the accordance of architecture and landscape, it aims to create interesting and artistic landscape by reconstruct mountain and streams in geometry.

The design characteristics: the art of mother earth and the landscape without boundary.

特色标识小品

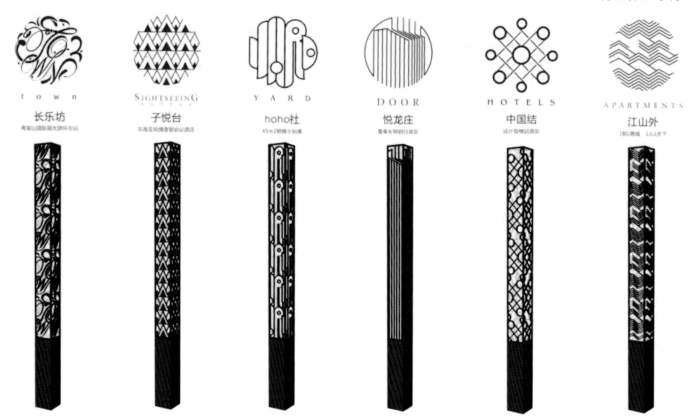

项目概况 Project Introduction

泰达·上青城项目总占地526余亩，由成都泰达新城建设发展有限公司集合中国、美国、德国、法国、菲律宾、新加坡等六个国家的建筑规划景观设计大师历时两年多，精心雕琢、倾力打造的，融合了泛亚洲度假建筑、现代中式建筑、川西民居建筑精髓的，集酒店、商业、会议于一体的信息化、生态型综合配套居住区。

形象样板区的设计通过对当地的自然特征和秀美的自然风貌的理解，从"山的印象"主题售楼部上吸取灵感，呼应现代感的建筑造型，采用解构的方式重新定义"山的印象"主题。建筑与景观相映成趣，融为一体，设计出具有艺术气质的景观环境。

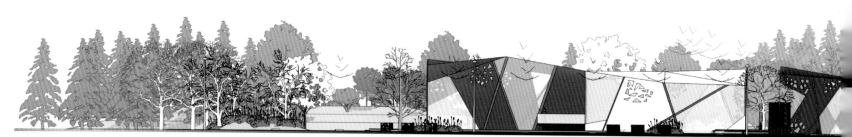

■ 会所与周边环境立面图　Elevation Drawing of the Club and Surrounding Environment

■ 特色图案式铺装　Feature Pattern Decoration

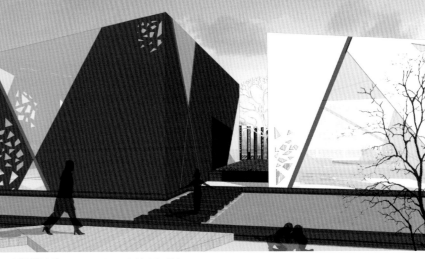

■ 会所后场水景　Water Landscape behind the Club

The project takes 526 Mu and is a 2-year masterpiece of landscape designers from China, US, Germany, France, Philippines, Singapore etc. It is managed by Taida Ltd, and is a combination of Asian, modern Chinese, and West Sichuan resident styles. It serves as a refreshing and comfortable hotel, and is with commercial and meeting functions.

The sample area design is based on the comprehension of local natural characteristics and fascinating sceneries.

The inspiration is also derived from the "The mountain Impression" sales department. To go in line with its modern architecture, it redefines the theme by deconstruction and form the artistic work of landscape.

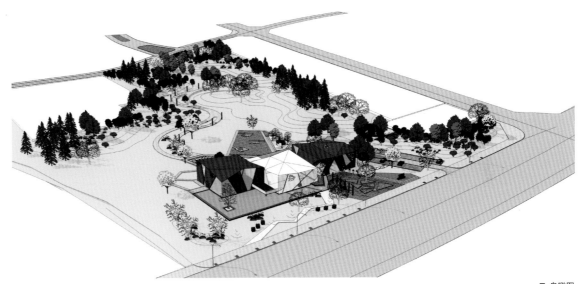

■ 鸟瞰图
Aerial View Drawing

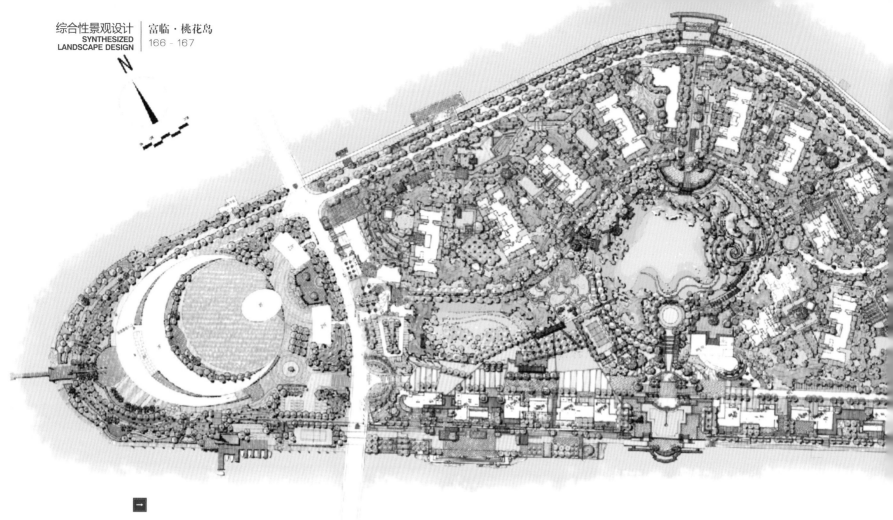

分享之岛，水上天堂
富临·桃花岛

The Island for Sharing and the
Heaven Above Streams

Fulin，Peach Flower Island

项目开发商： 四川绵阳富临房地产开发有限公司

项目地址： 成都绵阳市高新区三江交汇处湖心岛

设计面积： 470余亩

设计单位： 成都绿茵景园 设计一所

方案主创： 潘旭、张彬、郭伟、孙斐

方案团队： 汪雅兰、李茂、陈辉

合作单位： 新西兰LITTORALIS设计公司

施工图团队： 阙龙庆、姚舒雅、曾繁利、潘宇、陈浩然

发展阶段： 2008-2009年景观规划设计

完成阶段： 2010年完成一期滨江带商业街景观施工图和施工

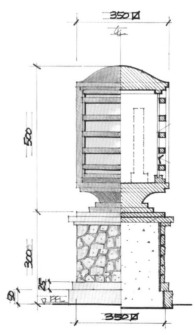

②- 特色灯柱详图-A

项目概况 Project Introduction

"桃花岛"位于四川省第二大城市绵阳市，距离省会成都120公里，地处绵阳三江交汇口，其周围水域面积达5.05平方公里。北望延绵起伏的富乐山，西南临风景秀丽的南山公园，西北靠中继岛生态风景区，东南方则是宽广的培江水域。拥有中国罕见城市湖心岛，岛内视野开阔，景色优美，地理环境优越，生态环境良好。岛头与其西面的中继岛生态栖息地仅百米之遥，与其西南面的南山也只有一江之隔，其生态重要性不言而喻。岛尾东北面的沿江区域与江对岸的富乐山生态保护区直接对应，因此这段河岸的景观设计也应强调生态恢复与保护。

The peach island is located in Mianyang, the second city of Sichuan, 120km away from Chengdu and at the crossing of 3 rivers. Its water area covers 5.05 square km. The undulating Fule mountain is adjacent to its north, South mountain park to the southwest, Zhongji eco scenic area to the northwest and Peijiang river to the southeast. It prides itself on being a rare urban island in rivers with broad horizons, breathtaking scenery, superior geographical location and excellent eco environment. Opposite to the Fukeshang eco conservation, it also place emphasis on the ecological restoration and protection.

发展规划理解 Onderstanding the Planning

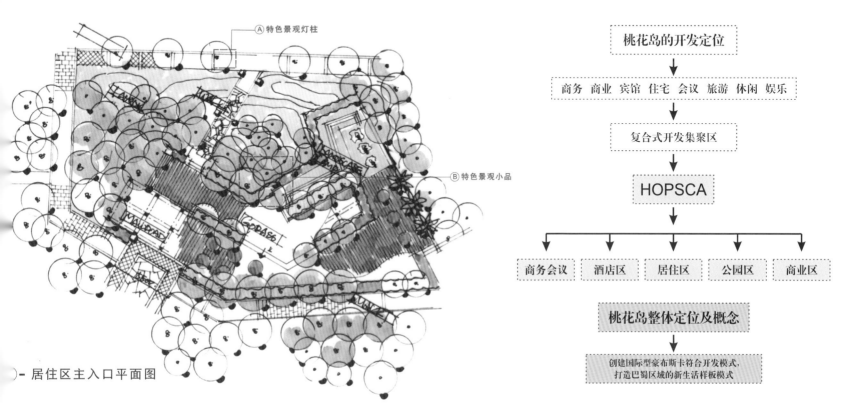

- 居住区主入口平面图

综合性景观设计 | 富临·桃花岛
SYNTHESIZED LANDSCAPE DESIGN

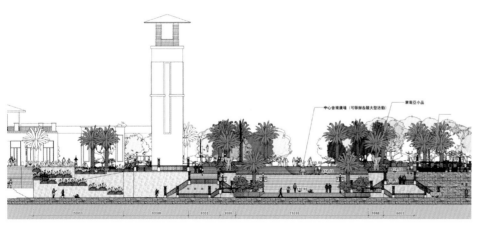
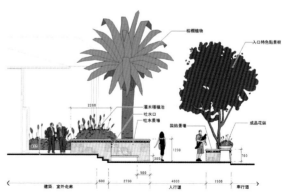

设计构思 Design Conception

在对"桃花岛"方案设计中，针对它的特殊性进行规划时，是通过对城市结构、区域交通、区域商业、区域生态结构在岛内周边关系以及岛内自然条件等方面全面分析得出的结果，并运用景观规划设计专业的国际先进理念，集中解决了排水、功能布局、交通结构以及框架等植物景观规划设计中的关键性问题。

The design attaches importance to its particularity of the project, after all-sided analyses of urban structure, regional traffic, regional commerce, regional eco-structure, and the natural condition of the island. We employ advanced international conceptions to solve the problems of drainage, functional arrangement, traffic structure and frame.

■ 桥上商业区域的视觉感受　Visual Feelings in the Business Region on the Bridge

■ 商业区的表演竞技场
Performance Arena in the Business Region

■ 临江商业步行街
Riverside Business Street

■ 表演竞技场夜景
Scenic Sight of Performance Arena

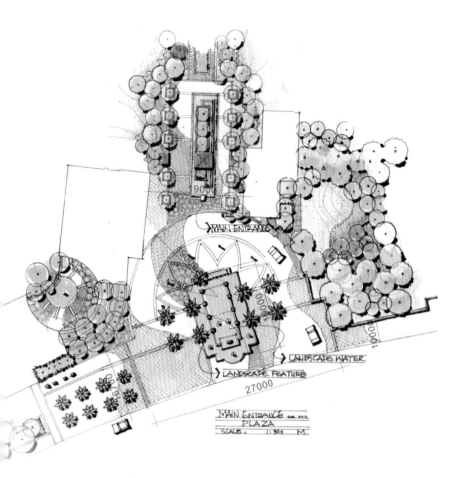

首先就形态结构来说，相比美国的纽约、丹麦的哥本哈根等城市来说，绵阳的城市脉络体现出了尊重自然的城市发展原则，人为规划的痕迹较轻。所以桃花岛的开发将形成一个交通环线，加强桃花岛与现有商业中心的连接。岛内集商业、宾馆、会议、办公、居住、休闲旅游、运动、公园为一体的复合式开发模式。在三江汇合口促成一个新的城市CBD并与现有城市商业中心遥相呼应，推动城南新城发展。

其次场地周边的三江汇合口的生态环境和亲水感受为城市休闲空间营造提供了得天独厚的的自然条件。加之现有的滨江绿化带和公共绿地已初步形成了受市民欢迎的休闲空间。它的岛头位置与绵阳城区及两岸山体上的富乐阁和南山塔等历史建筑的遥相呼应，既增加了新的亮点，又与其西面的中继岛生态栖息地仅百米之遥，与南山公园也只隔一江之水。所以桃花岛的开发无疑带动了周边商业发展。

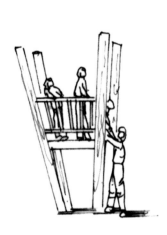

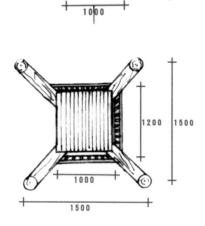

■ 示意图 Reference Drawing

■ 示意图 Reference Drawing

■ 趣味儿童设施
Children Facilities with Pleasure

综合性景观设计 | 富临·桃花岛

居住区详图

居住区组团平面图

ⓐ - 特色景观墙
ⓑ - 特色景观墙
ⓓ - 坐凳挡土墙

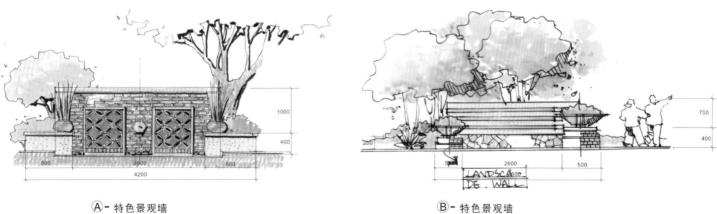

To begin with, compared with New York, Copenhagen and cities of this kind, and in terms of structure, the city shows the respect to the nature and is with less artificial planning. The development of Peach Island will form a traffic circuit and strengthen the connection between the island and the commercial center. The island development embodies commerce, hotels, meeting, office work, living, restful traveling, sports and park functions. Echoing with the downtown commercial center, a new city CBD will be constructed at the river crossing to promote the south development.

In addition, the river crossing and the greenbelt beside river provide restful space for citizens. And its adjacency to the city and historical architecture add color to its charm. Being in vicinity to the Zhongjidao island and South park, its development brings about the neighboring commercial development.

三、植物框架上

1、应该重点对岛头的西侧及岛内东北沿岸进行改造和生态恢复。

2、由于生态区域的重要性，应尽可能的在其他区域营造生态价值高的绿地空间，并在岛内北岸和中心区域创造生态走廊。

3、其他区域植物以景观性为主要目的，考虑视觉景观效应，景观空间尺度控制，以及微观气候调控。

The Vegetation Frame

1.It is supposed to be focusing on the restoration and redevelopment of the west side of the island head and northeast bank.

2.Due to its eco-value, green space are supposed to be built in other regions and eco-corridor is to built.

3.The plants in other regions are arranged for its visual scenic effect, the landscape space control and micro climate control.

最后是框架概念的设计

一、生态排水系统

1、根据自然地形以及道路规划标高，分为三个集水区，并结合生态导水渠沿道路将雨水导向三个指定点。

2、中心区域地势较低，为生态湿地提供了基本条件。

3、岛头和岛尾集水区相对面积小，在排水口设置小面积的生态净水植物池便可有效净水，然后直接放入河道。

Last but not the least, about its frame concept design:

The Eco-drainage System

1.The island is divided into 3 water collection areas according to its natural landform and road planning, to which rain water is led to by eco-headrace.

2.The central area is low enough to provide condition for eco-everglade.

3.Comparatively, the head and tail part of the island are small. Little size eco plant pools are arranged at the rainspouts to let the water into rivers after purification.

二、功能与交通结构

1、主要将公共活动区域集中在岛内会议中心-步行街-酒店公寓轴线上，尽量避免公共区域与居住区的冲突，减少人类活动对生态价值较高区域的干扰。

2、在沿岸生态及景观价值高的区域设置环岛步行及自行车道，为游客、市民和岛上居民慢跑步等健身方式增加有利条件。

3、设置一系列亲水休闲活动和水上运动设施，包括桥西侧的水上运动俱乐部，桥东侧的室外咖啡、茶室等。

The Functional and Traffic Structure

1.Public activity areas concentrate on the axis of meeting center-pedestrian street-hotel to avoid the conflict with the residential area and interference with eco-area.

2.Walking tour and bicycle lane are arranged for visitors and citizens working out in the high value landscape area.

3.A series of restful activities beside the river and water sports facilities including the club and outdoor cafes, teahouses are laid out.

综合性景观设计 | 富临·桃花岛

→ 主题定位 Theme Orientatin

奢华享受的度假式会议中心+异域风情商业街+
高档住宅+生态休闲自然式公园

度假气氛+绵阳会客厅

↓

城在水上，林在城中

分享之岛 水上天堂

↓

新生态，新人文，新生活

循环能源、无污染、可循环系统、场地精神、风水

景观效应，生态治疗

人文和景观风格相互渗透的无界景致

■ 迎宾中心广场 Hosting Center Square

→ 设计理念 The Design Conception

在对它前期分析之后，在下面具体的方案规划设计中把它主要设计规划为会客厅、风情商业街、呼吸公园、会所泳池、湿地公园、森林小区、驳岸这几个主要区域，并且使用了新颖独特的且可行的雨水收集方式。

会客厅：顶级会馆，以时尚崭新的现代风格，张扬科技发展观与创造性规划理念，宽敞壮丽的会客广场，融入地域文化形成展示窗口，豪华游艇俱乐部凸显贵族新文化，力求将会客厅高端品质演绎到极致。

风情商业街：以独特的巴厘岛休闲异域文化打造度假氛围，"订制"营造生态个性商业。商业前庭后院利用驳岸的高差形成错落变化的商业空间，局部以亲水台阶结合生态湿地软化驳岸，在满足防洪需求的同时，亦增加了景观的情趣。巧妙利用了防洪堤扩展了商业空间，增加更多形式和风格，体现出文家的包容性。设计融入居家花园的模式打造商业街区，打破了商业空间大量试用硬质铺装的传统，把生态的概念具体化，使之形成个性十足、休闲风味更浓的商业街区。

呼吸公园：以一种回归自然的设计理念，用思维去创造轻松、自然、富有情趣和探索性的开放性绿地。以简洁省材的手法营造自然、灵动的空间，文脉通过运动交流、游玩、思索等方式自然而然侵入人的心里。

穿梭在起伏幽深的白桦林中，漫步在蜿蜒的木栈道上，两半的野花，放眼的大草坪，婀娜多姿的白桦树，构成一幅纯净、梦幻得让人无尽陶醉的梦的天堂。

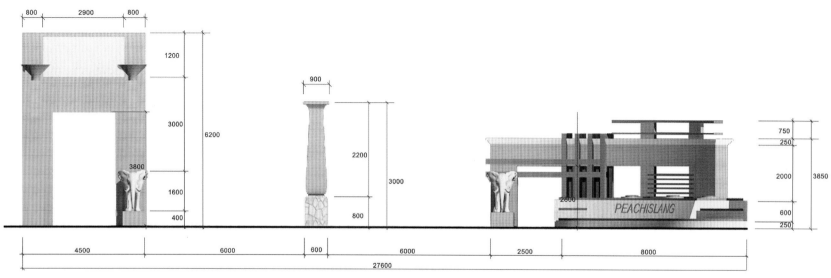

桥头入口构架立面图-1

公园内不提倡机械化、模式化的游乐场所。脚下的土地、周围的草坡、可利用的材料等即可激发无限可能。尽量用自然的方式让人们去寻找、去探求，以自己的方式玩出个性花样。包括在触觉、听觉、视觉、味觉等方面唤起人们对大自然的兴趣，从而更好的学习自然，尊重自然。

湿地公园：通过天然雨水收集过滤系统，使岛内水源能通过自然渗透、过滤，进行循环利用。

森林小区：以开放的空间形成岛内特有的生活自由圈。公园作为附属的公共产物，也是小区的舒适前厅。小区内空间作为其后花园，以绿色最大化的设计理念，减少后花园的铺装空间设置，使运动及功能性的空间往前厅转移，让功能空间得到最大化的利用。节约了用地，同时也为小区森林化提供更多的可能。

驳岸：最具有价值的一条纽带，视线的窗口，和培江有着最直接的联系，景观效果的好坏直接影响岛内整体形象。

■ 桥中 Middle of the Bridge

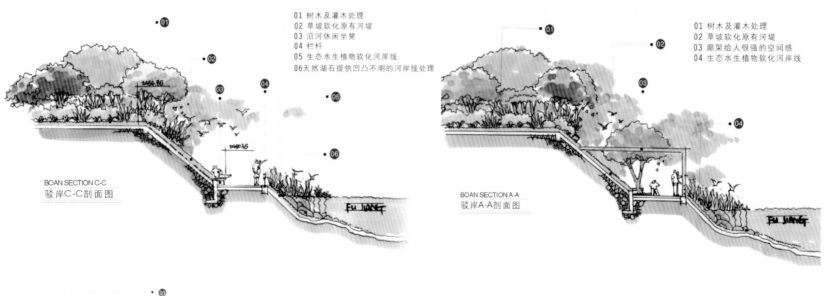

BOAN SECTION C-C
驳岸C-C剖面图

BOAN SECTION A-A
驳岸A-A剖面图

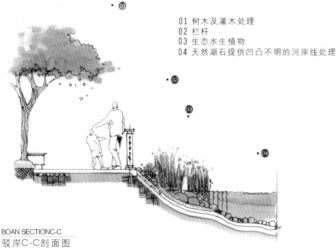

BOAN SECTION C-C
驳岸C-C剖面图

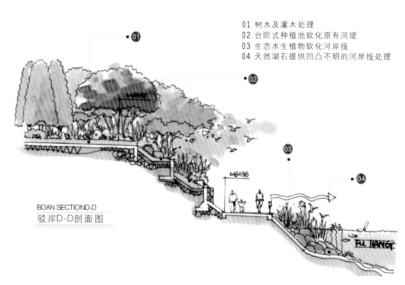

BOAN SECTION D-D
驳岸D-D剖面图

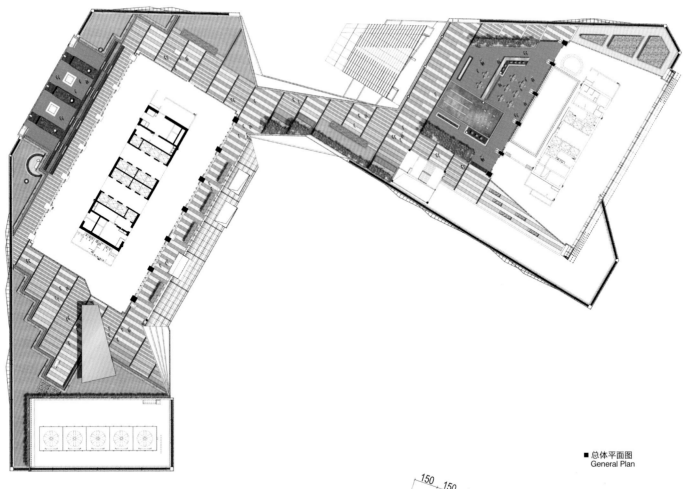

■ 总体平面图
General Plan

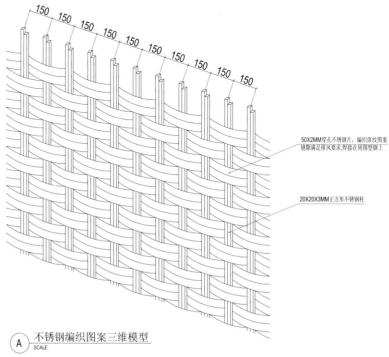

A 不锈钢编织图案三维模型

仁恒置地广场
Renheng Zidi Square

开发商：仁恒置业（成都）有限公司
项目地址：人民路南路二段1号
总建筑面积：190980平方米
规划的办公楼建筑面积：72073平方米
商场建筑面积：64556平方米
酒店式公寓的规划总建筑面积：51415平方米
项目类型：甲级写字楼、酒店服务式公寓及国际品牌购物中心
合作单位：EDAW景观设计公司
景观施工图设计：成都绿茵景园 设计三所
主创团队：曾健、潘强、潘宇、雷冬
设计时间：2008年

■ 造型种植花坛透视手绘效果图
Manual Perspective Effect Drawing of Flower Terrace for Modeling Planting

设计主题 Design Theme

仁恒置地广场，位于成都城市中轴线人民南路二段1号，地处城市CBD核心区，承续城市人文脉络，集萃高端商务商业精华，共享城市繁荣发展，辉映蓉城锦绣风光。

仁恒置地广场，成都CBD核心的高端建筑综合体，中国指数研究院评选为"2008-2009中国城市地标建筑"，汇集国际品牌购物中心、超甲级写字楼、酒店式服务公寓等三种高端物业形态。立足国际视野，世界标准，以前瞻性的定位、领先的建筑硬件、国际水准的运营管理，建筑世界级的商务/商业平台；以国际奢侈品、跨国企业、全球高级商务人士的聚合，成就中国西部经济地标、时尚中心。

成都国际品牌聚集地，营造奢华时尚体验

仁恒置地购物中心，汇集世界知名品牌形象旗舰店、国际高档品牌服饰专卖店、各种主题概念店/体验店、高档餐饮休闲等，以品牌优势保障高端定位；以对时尚的敏锐洞察，与世界潮流保持同步；以多功能多业态的组合，营造超凡的高端购物体验和愉悦精神享受；以国际品牌的文化内涵和时代气息，确立全球奢华时尚之旅的中国西部坐标。

超甲级写字楼，跨国公司区域总部

仁恒置地广场，国际标准超甲级写字楼，集合世界领先的建筑科技、一流水准的物业管理和完善的商务配套，提供宽敞舒适、高效环保的商务办公环境，充分满足全球化时代背景下，跨国企业、国际机构和中外知名公司对商务办公的高标准要求，协助入驻企业提升形象与绩效，成就其在中国西部的区域发展总部。

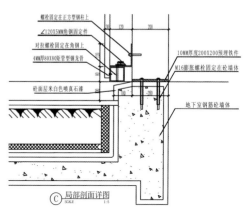

Ⓒ 局部剖面详图

■ 水景透视效果图
Perspective Effect Drawing of Water Landscape

■ 建筑灰空间透视效果图
Perspective Effect Drawing of Transitional Space of Architecture

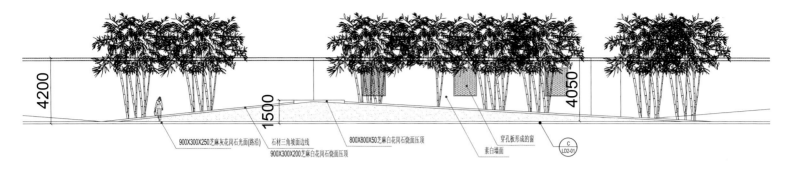

B 坡面B立面图 SCALE 1:150

"辉盛阁"酒店服务公寓，国际商旅人士五星级的家

仁恒置地广场"辉盛阁"酒店式服务公寓由享有盛誉的新加坡辉盛国际管理公司管理，"辉盛阁"为辉盛国际旗下高端品牌，以金质的产品和服务，提供精致、高雅、品味的居家生活。

仁恒置地广场"辉盛阁"公寓，位于成都红照壁高端精品商务商业区，周边CBD国际机构云集，各类购物商家及餐饮资源汇聚，成都风景名胜及文化体育场馆等环伺；交通便利，距离成都双流国际机场仅约20分钟车程。

仁恒置地广场"辉盛阁"提供360间豪华套房，面积从90-360平方米不等。室内装修由香港知名室内设计师梁志天主理，格调华贵，富于品位。同时，按五星级酒店标准，配设大堂、餐厅、恒温泳池、健身房等。以优质的硬件配置与"辉盛"体贴入微的专业服务，让国际高端商务人士、休闲旅客享受到舒适惬意的高尚生活。

■ 广场局部景观透视图 Perspective Effect Drawing of Sectional Landscape of the Square

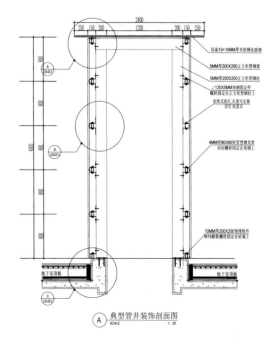

A 典型管井装饰剖面图 SCALE 1:20

■ 入口空间透视效果图　Perspective Effect Drawing of Entrance Space

设计理念　Design concept

采撷"天府之国"山水文脉，创造城市独特标识

仁恒置地广场，以成都得天独厚的城市人文氛围与其丰富的周边环境因素为灵感，采撷"天府之国"山水地域文化因子，与现代建筑的简洁明朗相结合，充分展现出高新科技表现主义与深厚人文主义的融合。建筑汇集的所有元素，均表达了同一个主题：唤起人类建立与自然的亲密关系，用建筑创造城市的独特标识。

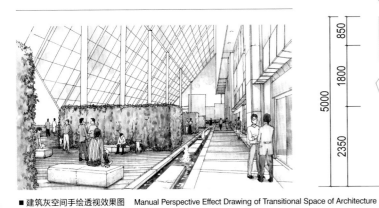

■ 建筑灰空间手绘透视效果图　Manual Perspective Effect Drawing of Transitional Space of Architecture

城市公园绿地景观设计
Urban Park Greenland Landscape Design

市民社会融合载体

　　城市公园是城市美化、改善生态环境的重要载体，同时也可以成为城市旅游中心。现代公园，不再是少数人所赏玩的奢侈品，而是普通公众身心愉悦的空间。城市经济水平的提高使人们越来越多的关注及参与到城市公园的建设过程中。在本次展示的公园绿地项目中，有的是综合性公园；有的是主体性公园；有的是社区绿地……每一个项目，绿茵团队都会整合最优势的力量进行反复研讨，力争可以重新组织构建城市的景观，组合文化、历史、休闲的要素，使城市重新焕发活力。

城市公园绿地景观设计
Urban Park Greenland Landscape Design

182	188	196	200	206	212
汶川·水磨镇5.12灾后重建景观工程	青城山世界自然遗产灾后生态景观恢复工程	重庆市茶园新区假日公园	重庆市江津区滨江新城公园	重庆市江津区琅山大道公园	重庆市九龙坡区西彭组团J分区休闲公园
Wenchuan,shuimo Town Landscape Reconstruction Project After 5.12 Earthquake	Ecological Landscape Restoration Project Of Qingcheng Mountain World Natural Heritage After The Earthquake	Chayuan New City Holiday Park In Chongqing City	Binjiang Xincheng Greenland Park In Jiangjin District,chongqing	Langshan Avenue Park In Jianjin Distric In Chongqing City	Jiulongpo District In Chongqing City Leisure Park In Division J Of Xipeng Group

废墟中托起希望，城镇的跃迁
汶川·水磨镇5.12灾后重建景观工程

Hope Rising in the Remains, Transition of the Town
Wenchuan—Shuimo Town Landscape
Reconstruction Project after 5.12 Earthquake

援建单位：佛山市对口支援地震灾区水磨镇恢复重建项目代建管理中心
地　　点：汶川水磨镇
设计面积：90000平米
合作单位：北京大学中国城市设计研究中心
设计单位：成都绿茵景园 设计三所
项目负责人：肖浩波
设计团队：杜佩娜、潘强、彭溟鸥、马卉、袁怡、陈浩然、粟凡粒
发展阶段：2009年景观设计
完成阶段：施工已完成

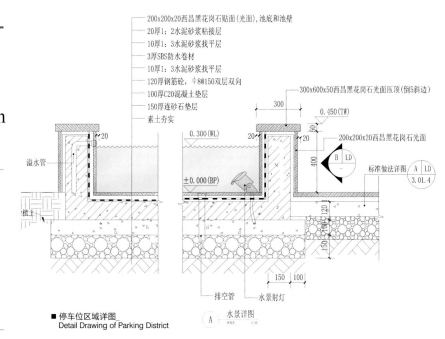

■ 停车位区域详图
Detail Drawing of Parking District

设计主题 Design Theme

"谷口莺啼细竹,洞门犬吠桃花。驻世何须丹灶,仙风吹长灵芽。"

At the entrance of the valley, the warbler is singing in the bamboos and at the gate of the cave, the dog is barking to the peach flowers. Without the eli-xir of life, one can stay young forever here, because there is a magic bud growing in the fairy wind.

明代的古戏台、清代的大夫第……这座商代碑文里就有记载的古镇阿坝州的"阳朔西街"、"凤凰古城",这就是千年古镇长卷如画的水磨镇。进驻水磨伊始,设计团队就明确了重建思路——造一座汶川生态新城、西羌文化名镇。

This is Shuimo Town like a large picture with a history of thousands of years. There are ancient drama stages in Ming Dynasty and officials' mansions in Qing Dynasty. There is a record in the inscription on a tablet in Shang Dynasty about the construction of "West Street of Yangshuo" in Aba Prefecture and the construction of "Ancient City of Phoenix" in Aba Prefecture. When entering Shuimo Town, the design team made a clear reconstruction aim, that is, to construct a new ecological city in Wenchuan and a famous town with Western Qiang culture.

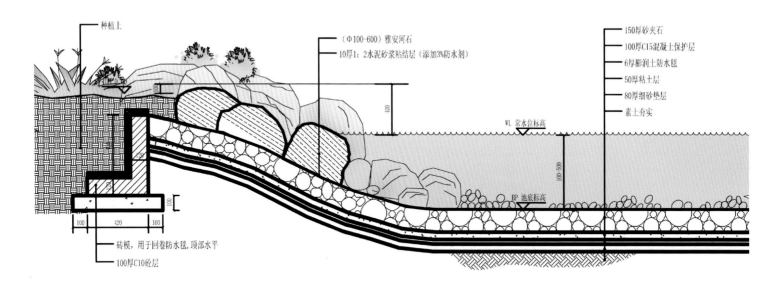

■ 驳岸详图1
Detail Drawing One of Bank Revetment

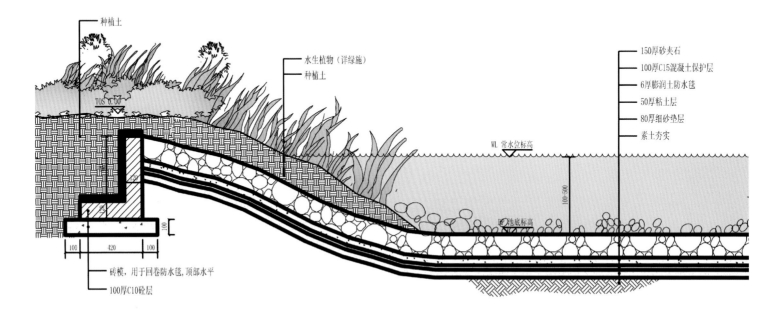

■ 驳岸详图2
Detail Drawing Two of Bank Revetment

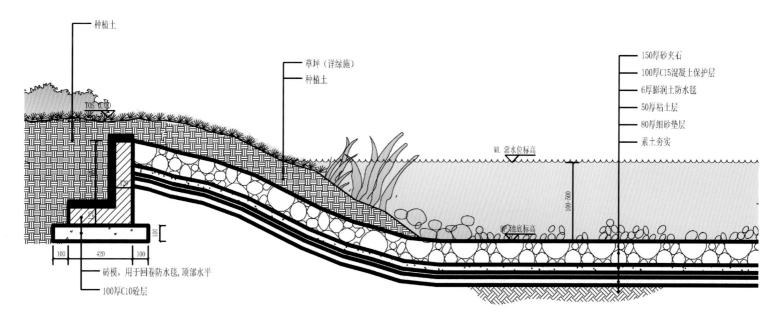

■ 驳岸详图3
Detail Drawing Three of Bank Revetment

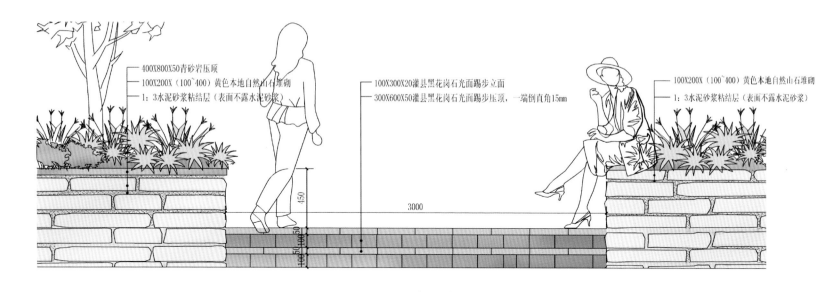

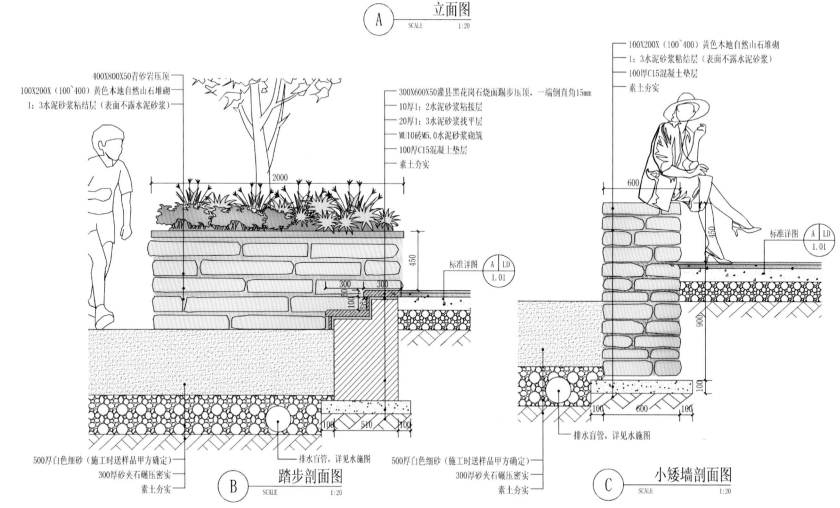

■ 区沙坑区域详图
Detail Drawing of Sandpit

景观设计不仅仅是一种操作，而是对现实环境的一种关照，我们相信人与人之间的互动，将赋予空间更深层的感动。因此我们在进行规划设计时，更多是考虑"以人为本"的设计宗旨，以整体优先、生态优先的原则，力图营造一个人工环境与自然环境和谐共存，以舒适温馨为特征的多样化空间体系。

本方案在景观设计上依托汶川水磨地区独特的藏羌文化特色，充分利用其优越的河流及生态资源，通过自然与人工的高度融合、人文景观的交织、空间景点的多层变化，追求飘逸自然风韵，在完善市政滨河公园功能的前提下，着力于"因地制宜""景到随机"，通过"屏俗收佳"等造景手法，剪辑、调度、点缀环境，"自成天然之趣，不烦人事之工"，创造出以天然景观为主，人工造景为辅的园林滨江公园景观画卷，营造优美、温馨、舒适的居住环境，为水磨人民提供一处优美舒适的现代城市公共休闲场所。

Landscape design is not only an operation, but also concern for the actual environment. We believe that the interaction between people can give space a deeply touching sense. Therefore, when we carry on planning and design, we pay more attention to the design aim of "people oriented" and the principle of overall priority and ecological priority. We try to create a diverse space system which is comfortable and warm, with a harmonious coexistence of artificial environment and natural environment.

In the design of landscape, the plan is based on the unique Tibetan-Qiang culture in Wenchuan Shuimo Area and makes a full use of its advantageous rivers and ecological resources. The plan seeks the elegant natural charm through the integration of nature and artificiality, the interweaving of man-made landscapes and changes in levels of space spots. On the premise of improving the function of Municipal Riverside Park, the plan puts emphasis on "adjusting measures to local conditions" and "creating the landscape at random" to clip, schedule and decorate the environment with the method of "abandoning the vulgarity and absorbing the essence". There is the pleasure of making from nature which relieves people of boring work. Natural landscape is given priority and artificial landscape plays a complementary role in the landscape constructing of the riverside park. The plan is aimed at creating beautiful, warm and comfortable living environment to provide people in Shuimo Town with a beautiful and comfortable public leisure place in the modern city.

城市公园绿地景观设计 | 青城山景观恢复工程
URBAN PARK GREENLAND LANDSCAPE DESIGN | 188 - 189

青城山世界自然遗产灾后生态景观恢复工程

Ecological Landscape Restoration Project of Qingcheng Mountain World Natural Heritage After the Earthquake

项目名称：青城山新火车站周边景观规划设计
项目规模：28万平方米
设计单位：成都绿茵景园　设计二所
项目地点：四川省都江堰青城山
委托单位：都江堰市政府，成都建工集团
合作单位：高士尔国际（英国）设计顾问有限公司
主　创：阿笠、刘丽红
团　队：张洪源、刘鸠鸠、粟凡粒、雷冬、袁亚宁
设计时间：2010年

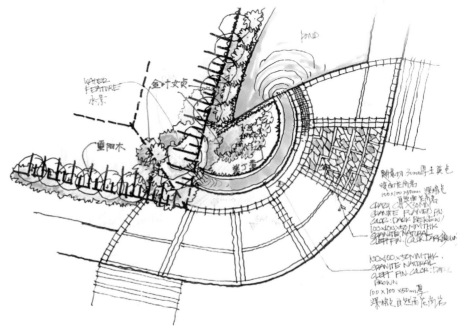

项目简介 Introduction to the Project

作为青城山"5.12"地震后新形象的功能需要，对青城山新火车站中心广场及自然景观通廊，青幽路延伸线，东软大道，彭青线节点进行景观改造设计，塑造新青城生活形象，展现青城山旅游区道教文化与景区特色。

项目位于新建成都至青城山景区高铁站出口处，是青城山景区重要的门户地段，与山门景区连接处重要的人行、车行与商业区域。景观设计上提出"注重不在于游览者眼睛所看到的，而在于自己的心情"，将景观规划范围概括为一中心三节点：以青城山门广场作为中心，与快铁站前广场，东软大道节点，彭青线节点及各连接通道作为景观规划的框架。设计理念上遵循青城道家文化特征，并引入中国山水画的意境，将中国古文化特征与现代的场地设置融合，展示山水道派风景。

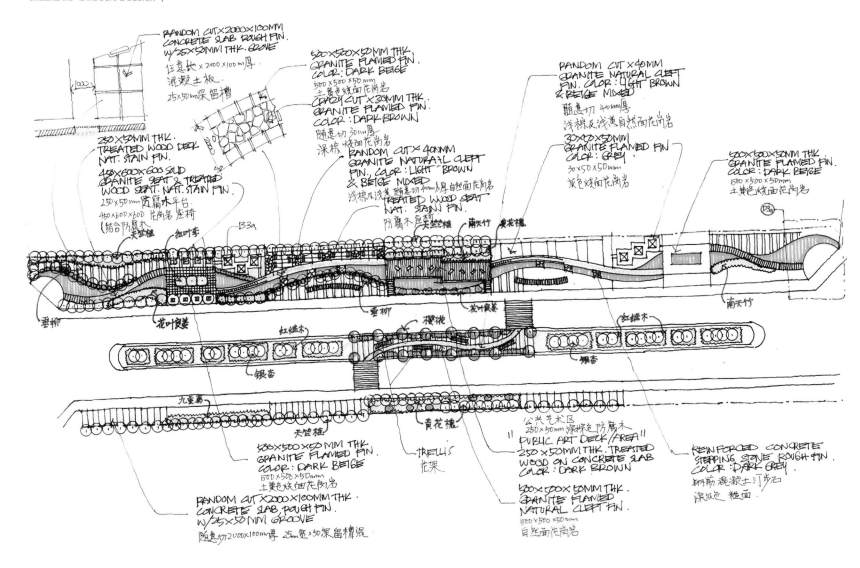

Introduction to the Project

To satisfy the function need of the new image of Qingcheng Mountain after "5.12" Earthquake, the landscape reform is designed for the Central Square in the new rail station of Qingcheng Mountain, the natural landscape corridor, the spreading line of Qingyou Road, Dongruan Avenue and joints of Pengqing Line. The design is aimed at shaping the new image of life in Qingcheng Mountain and presenting the Taoism culture in Qingcheng Mountain scenic region and the features of the scenic spot.

The project is located in the important section of Qingcheng Mountain scenic region and at the exit of the new-built high-speed rail from Chengdu to Qingcheng Mountain. There are important roads for pedestrians and vehicles and commercial regions connected with the scenic region. The design of the landscape doesn't emphasize what the tourists see, but what they feel. The planning scope of the landscape can be concluded into a center and three joints. That is the landscape planning layout with Qingcheng Mountain Entrance Square as a center and the square in front of the high-speed rail station, the joint of Dongruan Avenue, Pengqing Line joints and connecting roads as the joints. The design idea follows Taoist culture characteristics in Qingcheng Mountain and introduces the imaginary of Chinese landscape painting. It combines the features of ancient culture in China with the modern setting of the site to present the Taoist landscape.

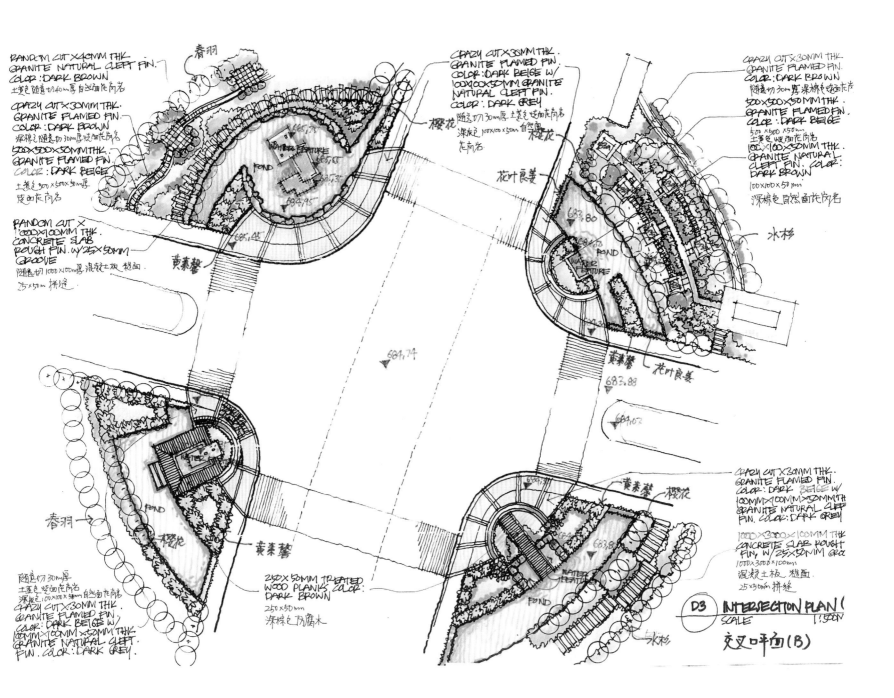

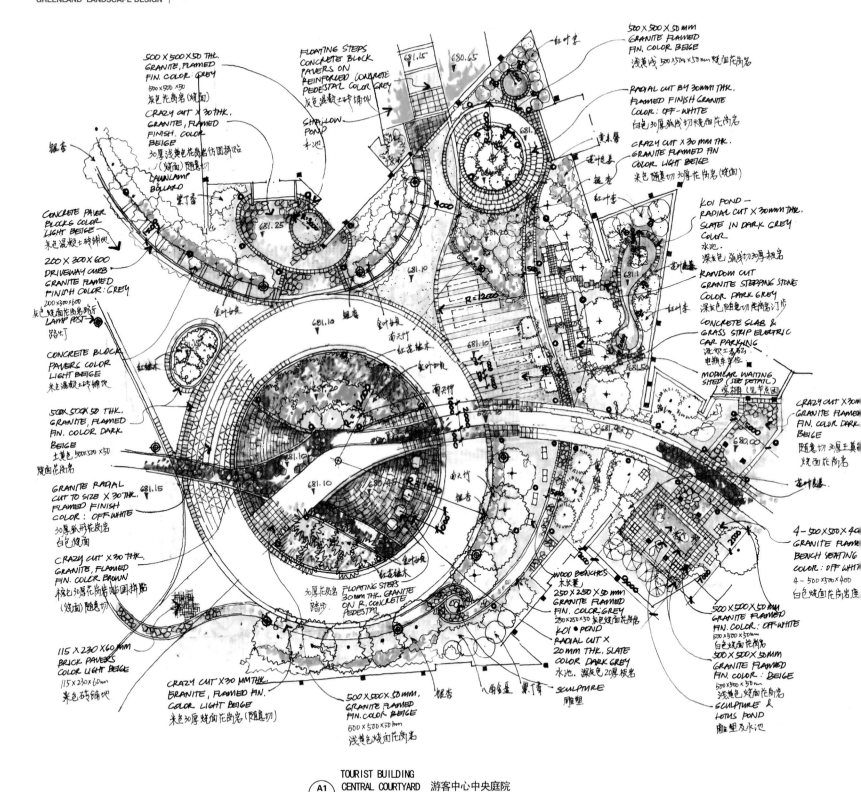

TOURIST BUILDING
CENTRAL COURTYARD 游客中心中央庭院

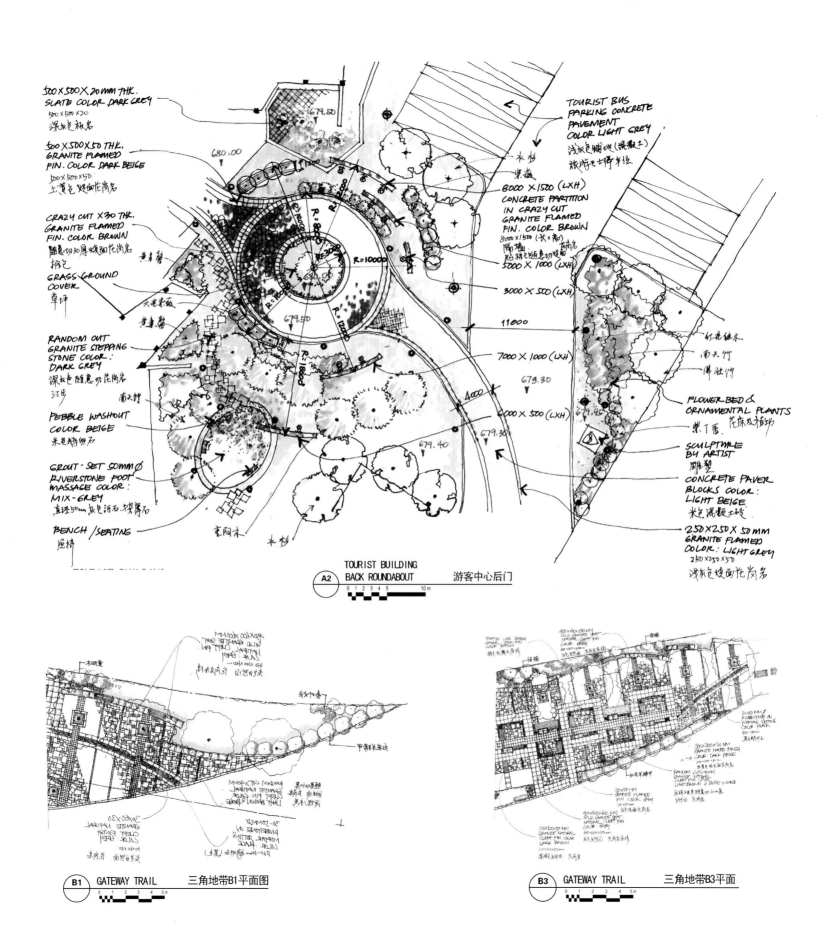

山水画 Landscape painting

中国古代的山水画通过特殊手法来创造精彩的抽象艺术，山水艺术的中心概念是描述蜿蜒小径，地与山之间的边界以及图中的主体视觉焦点。

山水与道教 Landscape and Taoism

道教认为"道"是宇宙万物的原本和主宰，也体现了道教、山水、青城山有着深层次的联系。道教中的自然与幽静是整个方案的概念。每一处视觉中心如山水画一般体现了道教的深层次的含义；抽象的意境让人们各自联想不同的感受。

山水画及视觉焦点 Landscape painting and visual focus

在整体规划中，道教的一些重要元素将体现在交通动线、特色景观及风景中，以创造出和谐的环境，同时营造出步移景迁的视觉效果。

设计方案减少公园大量的硬质景观，以还原山景原生态为主，通过流线型的道路与阵列式的林荫道，将人流引导至青城山门，并形成强烈的礼仪感与清幽感。同时在重要的各个节点上，将开敞的草地，图案化的水岸线，起伏的景观风景线三者结合，营造开放，自由，标识化的景区场地。通过植被的恢复性营造，以及新的功能场所与设施的布置，体现地震灾后青城山景区屹立不倒的精神形象，并持续散发出自身特有的道派自然文化。

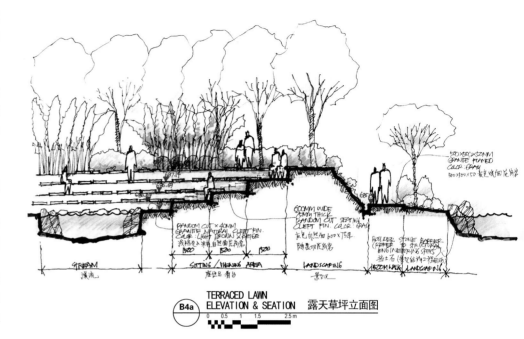

B4a TERRACED LAWN ELEVATION & SEATION 露天草坪立面图

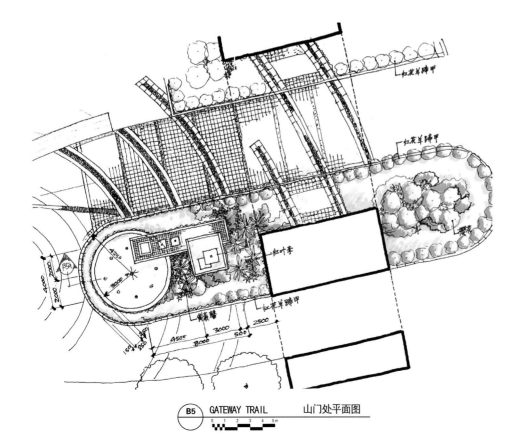

B5 GATEWAY TRAIL 山门处平面图

Landscape painting

The landscape painting in ancient China creates excellent abstract art with special methods. The central concept of the landscape art is describing the winding roads, the boundary between the land and the mountain and the main visual focus in the picture.

Landscape and Taoism

Taoism holds the opinion that "Tao" is the origin and master of everything in the world and also presents the relation in deep levels of Taoism, landscape and Qingcheng Mountain. Nature and quiet in Taoism is the conception of the overall planning. Every visual focus presents the deep meaning of Taoism and the abs-tract artistic imaginary makes people associate it with different feelings and meanings.

Landscape painting and visual focus

In the overall planning, some important elements of Taoism will be presented in the traffic line, featured landscape and scenery to create harmonious environment and to create the visual effect that landscape changes with the steps.

The design reduces the use of hard landscape and restores ecological mountain landscape primarily. It leads people to the gate of Qingcheng Mountain through a streamlied path and the avenue standing like an array and forms a strong sense of etiquette and quietness. Meanwhile, the combination of open grassland, designed coastline and undulating landscape lines at the important joints creates an open, free and symbolized scenic site. Through the creating to restore the vegetation cover and the arrangement of new functional sites, the design presents the firm mental image of Qingcheng Mountain scenic region and its Taoism culture Qingcheng Mountain continues to reveal.

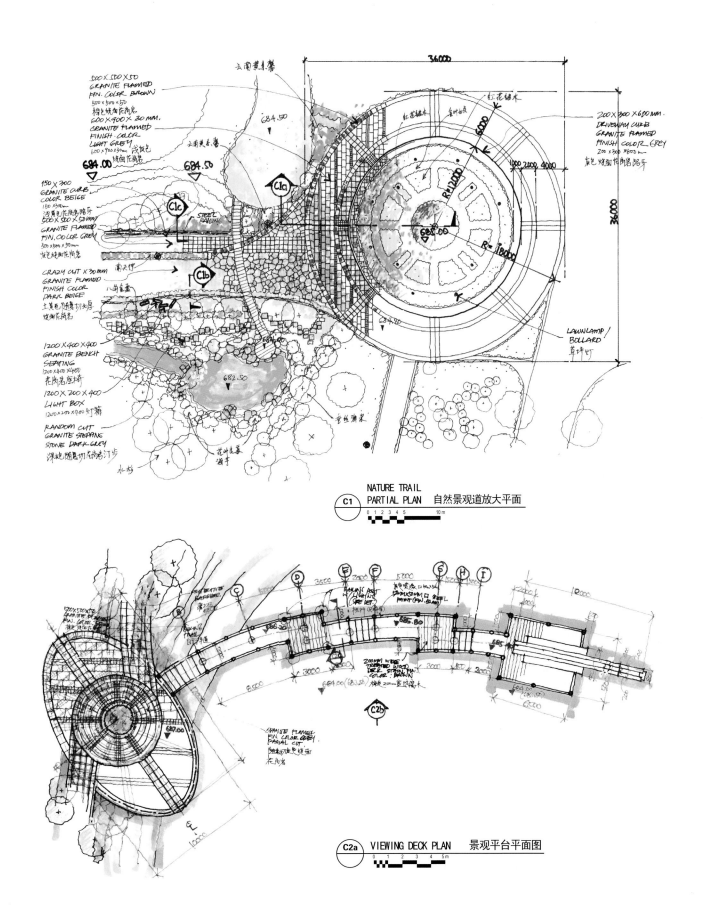

重庆市茶园新区假日公园
Chayuan New City Holiday Park in Chongqing City

项目类型：休闲娱乐城市公园
项目面积：28万平方米
设计单位：重庆蓝调国际（绿茵景园集团公司）设计一所
项目地点：重庆茶园新区
委托单位：重庆同景置业有限公司
主　创：任刚
团　队：毕领枫、樊菁、陈果、黄茜、何代军、李海平、张俊、汤远英
设计时间：2010年

■ 荷塘月色主题区效果　Effect of Lotus Pool by Moonlight Theme Area

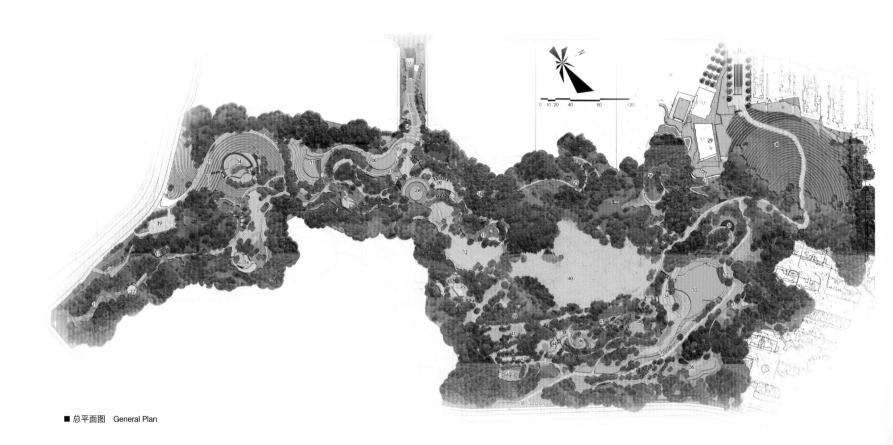

■ 总平面图　General Plan

■ 休闲游览区　Recreational Tour Area

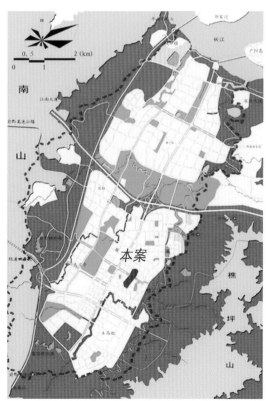

■ 项目区位图　Project Location Map

■ 项目简介 Introduction to the Project

　　重庆市茶园新区同景公园是在2010年重庆蓝调城市景观规划设计公司为同景国际城提供的设计方案。设计师对项目周边环境作了确切分析后对公园做了明确的定位。以茶文化为主题，运动休闲、生态教育为重心，提升茶园新区的城市形象，以此为核心向外辐射，吸引南岸区及主城各区公众参与，建成一个具有时代特色的大型城市文化运动公园。茶起源于中国，其精髓在于品茶、品道的哲学思想。

　　公园位置处于茶园新区城市景观轴线上，其关系是公园与城市，公园与新的区府中心的关系，就如同景观风水学中的穴与案山的关系。"凡案山，必横生、必特起、必有起有止、必身可凭、手可据、目可下视，方谓之真案山"。设计师提炼中国传统茶文化的哲学思想将山体公园打造成为绿树成荫的自然屏风，将公园入口广场打造成映衬自然的山水画卷，完美的体现了自然的风水格局，很好的协调了公园周边关系，并且能与轴线的区府相呼应。

　　Chayuan New City Holiday Park in Chongqing City is a design plan for Verakin New Park City provided by Chongqing City Blue Tone Urban Landscape Planning and Design Company in 2010. The designer made a definite orientation for the park after a careful analysis about the surrounding environment of the project. The image of Chayuan New City is promoted with a theme of tea culture and an emphasis of sports leisure and ecological education. Radiating from this central point and attracting people in the south bank area and main city, a large urban culture sports park with features of this age is constructed. Tea originated from China and its essence lies in philosophical thoughts of tasting tea and thinking about life.

　　The park is located in the urban landscape axis of Chayuan New City. The relation between the park and the city, and between the park and the new district center, is like the relation between the "point" and "slope" in the landscape geomancy. "A slope must be lofty and stand out. It must have a beginning and an ending. People can lie against it and hands can lean on it. People can look down on it. Thus, it is a real slope." The designer refines the philosophical thought of traditional tea culture in China and makes the mountain park a natural screen with green trees. The designer makes the entrance square of the park a landscape painting which presents perfectly the natural landscape layout. The square also coordinates the surroundings of the park and corresponds to the axis district.

城市公园绿地景观设计 | 重庆市茶园新区假日公园
URBAN PARK GREENLAND LANDSCAPE DESIGN

■ 禅茶品道景区　Scenic Region to Taste Tea and Realize the Philosophic Theory

设计通过改造部分原有市政道路，将其处理成景观化的入口空间。将公园的绿色延伸至城市内。设计采用了相对自然的手法，通过一道人行天桥，将人巧妙的引向公园内。同时也巧妙的运用了入口高差打造出富有气势的景观跌水。蜿蜒的路径仿佛是自然绿色的触角向城市生长，景观化的人行桥将成为公园标志的景点，在入口给人们留下深刻的印象，并激起人们的好奇心进入公园。半山的树阵广场给游人提供了休闲的活动空间。广场后面的山坡上一座以现代川东民居风格的茶文化馆在绿树间矗立，茶文化馆将集中的体现茶园新区的风貌和宜居重庆的景象，成为公园的核心建筑，茶文化馆外根据山地地形形成退台的观景休闲平台，将是人们露天休闲的场所。设计方案大量减少公园的硬质景观，以植物景观为主，还公园青山绿水的自然风貌。设计从不同的角度来表达了同景国际公园与城市的关系，将公园的功能活动及景观形态重新进行架构，所形成的茶园新区是一座富有人气和独特风貌且吸引人们前往的休闲娱乐的城市公园。

The design transforms some of the original municipal roads into landscape entrance space. The green of the garden spreads into the city. The design uses a relatively natural method to lead people into the park with a pedestrian bridge and also makes a good use of the height differences of the entrance. The design creates landscape drop water. The winding roads are like green feelers of nature growing to the city. The landscape of the pedestrian bridge will be the symbol of the park. It will impress people at the entrance and arouse their curiosity to enter the park. The Square of Tree Array located half-way down the hill provides tourists with leisurely activity space. On the hill behind the square, there is a Tea Culture Hall in a modern living style in the west of Sichuan standing in the green trees. The Tea Culture Hall will give a concentrative expression of the style and features of Chayuan New City, and the scene of Chongqing. The Tea Culture Hall will be the central architecture of the park. Outside the Tea Culture Hall, there is a sightseeing platform which is a terraced roof formed by the mountain terrain. It will be an open-air leisure place for people. The design reduces the use of hard landscape and mainly uses landscape of plants to restore the natural style of green mountains and clear water in the park. The design shows the relation between Verakin International Park and the city from different perspectives. It reforms the framework of the functional activities and landscape forms. The Chayuan New City is a city park for recreation with popularity and unique charm which attracts people to the park.

■ 绿地系统规划
Planning of Green Land System

重庆市江津区滨江新城绿地系统规划研究及三重点公园概念设计

The Planning Research of Binjiang New City Greenland System in Jiangjin District in Chongqing City and the Conception Design of Three Main Parks

项目类型：城市公共开放空间
项目面积：绿地系统30平方公里；三大重点公园约700公顷
设计单位：重庆蓝调国际（绿茵景园集团公司） 设计一、二、三所
项目地点：重庆市江津区滨江新城
委托单位：重庆市江津区规划局
主　创：张坪、张勇
团　队：任刚、王轶、敖翔、肖勇、陈云川、樊菁、毕领枫、陈果、杨佳、林凤君、李洋、黄茜、李锦香、熊阳梅、左春丽、张骊瑜
设计时间：2009年9月——2009年11月

燕子岩郊野公园总平面图
General Plan of Swallow Cliff Country Park

项目简介 Introduction to the Project

2007年9月20日，国务院正式批复同意《重庆市城乡总体规划（2007-2020）》，在最新一轮的总体规划中，确定了新的城镇体系结构，即：全市形成市域中心城市、区域性中心城市、次区域中心城市、中心镇和一般镇五个层次等级结构，（特大城市、大城市、中等城市、小城市和小城镇五个级别的规模结构）江津和万州、涪陵、合川、永川、长寿被定为六个区域性中心城市，在这样的历史背景下，江津积极调整与主城都市区的关系、拓展空间、修编城市总体规划。

本项目主要由两部分构成：滨江新城绿地系统规划及重点公园概念设计。滨江新城南临长江与几江城区隔江相望，西连缙云山脉，北临双福新区，东接九龙坡区西彭组团，规划面积30平方公里；重点公园包含燕子岩公园、中央公园及滨江公园三大部分。

经过前期大量现状及政策分析，本次绿地系统形成"青山环抱构筑生态屏障，燕子穿梭编制绿色网络"的布局结构模式，具体的讲即：利用景观生态学"基质"—"斑块"—"廊道"的基本原理，在自然山水的空间格局的基础上，建构合理的体系，体现"渗透、融合"的布局理念。

On September 20th, 2007, the State Council officially made a reply and approved of the "Overall Planning of Cities and Towns in Chongqing City（2007-2020）". In a new round of general planning, the new system structure of cities and towns has been defined. The new structure is that the overall city forms a fivelevel grade structure of municipal central city, regional central city, subregional central city, central town and general town and a fivelevel scale structure of extra large city, large, mediumsized city, small city and town. Jiangjin, Wanzhou, Fuling, Hechuan, Yongchuan and Changshou are defined as six regional central cities. In such a historical background, Jiangjin should adjust actively its relation with the urban area of the main city, spread space and revise the overall planning of the city.

The subject mainly consists of two parts: Planning of Binjiang New City Greenland System and the Conception Design of Important Parks. Binjiang New City is located along the Yangtze River in the south and faces Jijiang urban area over the Yangtze River. It connects with Jinyun Hill in the west, faces Shuangfu New Area in the north and connects with Xipeng Group in Jiulongpo District in the east. The planning area of the project is 30 square kilometers. The main parks include Swallow Rock Park, Central Park and Binjiang Park.

With abundant analysis of present condition and policy, there is a layout structure mode of "an ecological defense formed by the surrounding mountains, and a green net formed by the flying of the swallows in the Greenland System. Specifically, with the basic principle in landscape ecology of "matrix" to "patch" to "gallery", a reasonable system is constructed based on the space layout of the natural mountains and rivers. The system presents the layout concept of "penetration and integration".

马鞍石战场遗址效果图
Effect Drawing of Saddle Pebble Battlefield Relics

民俗文化作坊街效果图
Effect Drawing of Folk Culture Workshop Street

■ 都市之心夜景效果图　Effect Drawing of Heart of City Night Scene

三大公园主题定位为:
The Theme Orientation of Three Main Parks

燕子岩郊野公园 Swallow Rock Wild Park

体现岩石景观，保护现有的地形地貌、植被，保留排水体系，选择适当的植物（土层薄），安排活动（登山、眺望、果园、会议休假），考虑交通流线的关系（如何沟通、渗透融合）。

In the park, the rock landscape is presented, the present topography and vegetation cover are protected and the drainage system is retained. There are plants properly chosen (thin soil layer) and activities arranged (climbing the mountain, looking far at the scenery, orchard and conference vacation). Besides, the relation of traffic lines is taken into consideration (how to communicate, penetrate and integrate).

■ 欢乐艺术谷手绘效果（中央公园）
Manual Effect Drawing of Happy Art Valley (Central Park)

■ 五彩峡谷效果图（中央公园）
Effect Drawing of Colorful Valley (Central Park)

中央公园 Central Park

根据周围不同的用地条件进行功能分区，提出不同的景观意向理念，道路网络、人车停留节点（注意水面、高差处理），与城市其他公园绿地的沟通。

According to different land conditions around the park, there are function divisions and different landscape design concept proposed. There are also road nets, staying joints of people and vehicles (pay attention to the dealing of water and height differences) and the communication with other park green land.

■ 中央公园鸟瞰图　Aerial View Drawing of Central Park

■ 城市纪念公园手绘效果（中央公园）
Manual Effect Drawing of City Memorial Park (Central Park)

■ 花园演艺厅手绘效果（中央公园）
Manual Effect Drawing of Garden Performing Hall (Central Park)

■ 林荫广场（滨江公园） Shady Square (Riverside Park)

滨江公园 Binjiang Park

生态、景观，从城市向自然的过渡。

三大公园相互联系，互为依托，在滨江新城绿地系统规划的框架指导下，形成了"山环水绕绿荫满城，宜居宜业活力江津"的特色。

Ecology and landscape, the natural transition from the city.

Three main parks connect with each other and base on each other. Under the guidance of the framework of Binjiang New City Greenland System Planning, the feature of "the green city surrounded by the mountains and rivers and the active Jiangjin which is appropriate for living and working" is formed.

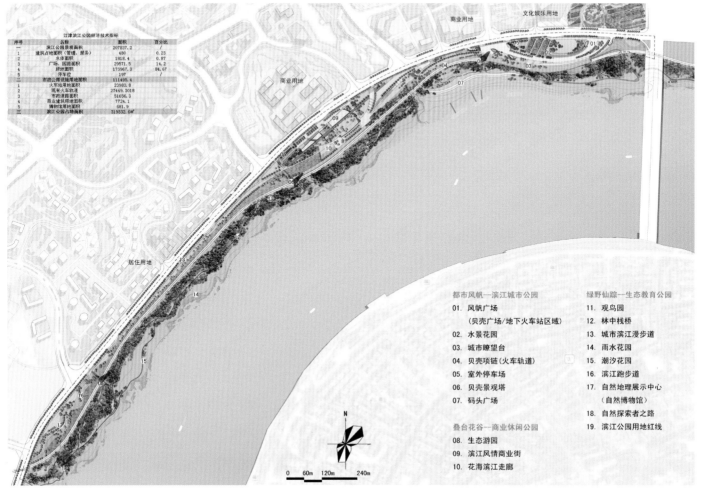

都市风帆—滨江城市公园
01. 风帆广场
　　（贝壳广场/地下火车站区域）
02. 水景花园
03. 城市瞭望台
04. 贝壳项链（火车轨道）
05. 室外停车场
06. 贝壳景观塔
07. 码头广场

叠台花谷—商业休闲公园
08. 生态游园
09. 滨江风情商业街
10. 花海滨江走廊

绿野仙踪—生态教育公园
11. 观鸟园
12. 林中栈桥
13. 城市滨江漫步道
14. 雨水花园
15. 潮汐花园
16. 滨江跑步道
17. 自然地理展示中心
　　（自然博物馆）
18. 自然探索者之路
19. 滨江公园用地红线

■ 滨江公园总平面 General Plan of Riverside Park

■ 都市风帆效果图（滨江公园） Effect Drawing of Urban Sail (Riverside Park)

■ 叠台花园效果图（滨江公园） Effect Drawing of the Garden (Riverside Park)

城市高压塔下的景观绿廊
重庆市江津区琅山大道公园

Landscape Green Gallery under the High Voltage Towels in the City
Langshan Avenue Park in Jiangjun District in Chongqing City

名　称：江津东部新城琅山大道公园景观设计
所在地：重庆江津区
投资额度：900万元
工期：2007.10—2008.12
设计公司：重庆蓝调国际（绿茵景园集团公司）设计二所
项目负责：王轶
设计团队：王轶、熊谊、吴东、潘启渝、艾珍、李海坪
规模：50000平方米
主要材料：花岗岩、青石
主要植物选择：黄葛树、银杏、天竺桂、杨树、杜鹃

■ 效果图　Effect Drawing

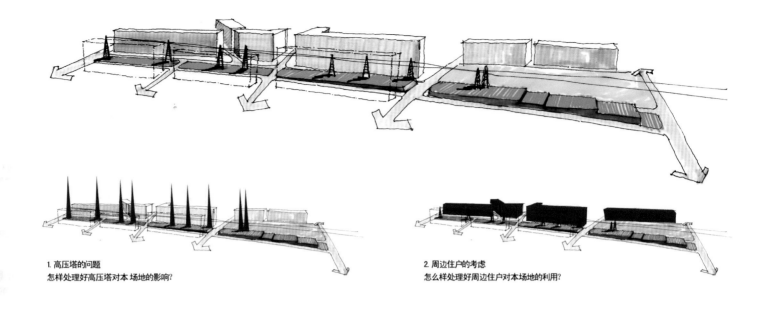

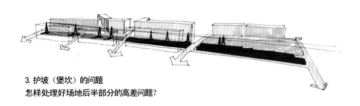

1. 高压塔的问题
怎样处理好高压塔对本场地的影响？

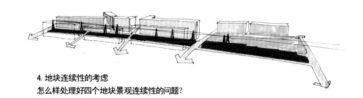

2. 周边住户的考虑
怎么样处理好周边住户对本场地的利用？

3. 护坡（堡坎）的问题
怎样处理好场地后半部分的高差问题？

4. 地块连续性的考虑
怎么样处理好四个地块景观连续性的问题？

■ 地块结构分析　Plot structure analysis

→ 设计思路　Design Idea

■ 手绘效果图　Manual Effect Drawing

琅山大道位于东部新城幸福大道和祥瑞商业步行街之间，起点为鼎山大道，终点为滨江东路。全长约1050米，宽约68米，其中中间景观带宽约44米，是东部新城规划的5条主要景观大道之一。

在项目初期，我们认真查勘了琅山大道的现状条件，仔细分析了场地内竖立的高压塔对人心理和身体上的影响。所以我们从这个角度展开反思，结合周边城市基础建设，来营造一系列以健康为主题的活动空间，倡导绿色健康的生活方式。而琅山大道是江津东部新城通往滨江"五纵"中重要的生态轴线，因此对其景观规划目标是：

1. 一条面向公众的生态景观带和城市景观干道。
2. 合理组织交通、功能、景观、绿化等，设置满足不同人群需求的服务设施，创造宜人的休息场所。
3. 消除高压塔对于周边住户的影响。
4. 利用场地地形构建丰富的山地景观
5. 恢复、创造多样性的自然生态环境。

在景观细节上我们利用当地的材料艺术化地强调和界定不同功能的休闲空间，旨在营造一系列的以健康为主题的活动空间来唤醒人们关注自身健康，为该区域的城市环境提供了一处非常珍贵的公共开放绿地空间。在中国城市建设继续朝着高密度的方向发展时，同存共居，健康生活成为琅山大道打造强劲理念。

城市公园绿地景观设计
URBAN PARK GREENLAND LANDSCAPE DESIGN

重庆市江津区琅山大道公园

Design Idea

Langshan Avenue is located between the eastern New City Happiness Avenue and Xiangrui Commercial Pedestrian Street, beginning from Dingshan Avenue and ending in Eastern Binjiang Road. The Avenue is 1050 meters long and 68 meters broad and the central landscape zone is 44 meters broad. Langshan Avenue is one of the five main landscape avenues in the design of New City.

■ 景观长廊 Landscape Corridor

At the early stage of the project, we made a careful survey of the conditions of Langshan Avenue. We also analyzed carefully the psychological and physical effect that the high voltage towers standing in the site have on people. Therefore, we developed our idea from this perspective and created a series of activity spaces with a theme of health to advocate the green and healthy life style. Langshan Avenue is an important ecological axis leading to Binjiang "Five Longitudinal Roads". Our planning goals of the landscape are as follows:

1. An ecological landscape zone and urban landscape road for the public.
2. Having a reasonable organization of traffic, function, landscape and greening, and serving facilities which can satisfy the need of different people.
3. Eliminating the effect of high voltage towers on surrounding residents.
4. Constructing various hill landscapes with the use of the terrain of the site.
5. Restoring and creating diverse natural ecological environment.

As for the landscape details, we use local material to stress and define artistically the leisure space with different functions. We aim to create a series of activity spaces with a theme of health to remind people to concern themselves with their health and provide a precious public green space for the urban environment in this area. Living in harmony and leading a healthy life create a strong concept support for Langshan Avenue when China continues to develop in the direction of high density.

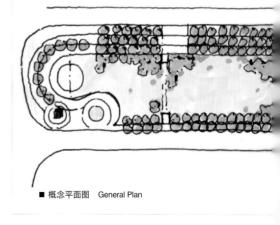

■ 概念平面图　General Plan

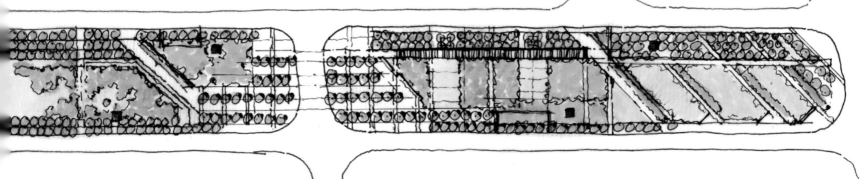

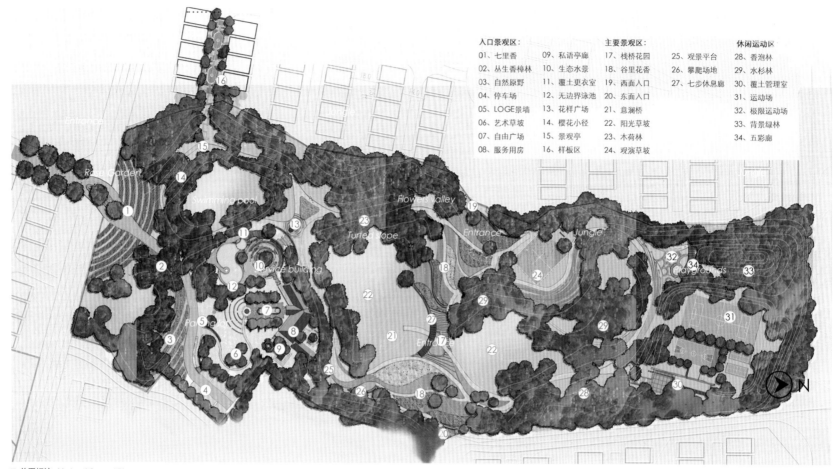

■ 总平标注　Marks of General Plan

入口景观区：
01、七里香
02、丛生香樟林
03、自然原野
04、停车场
05、LOGE景墙
06、艺术草坡
07、自由广场
08、服务用房
09、私语亭廊
10、生态水景
11、覆土更衣室
12、无边界泳池
13、花样广场
14、樱花小径
15、景观亭
16、样板区

主要景观区：
17、栈桥花园
18、谷里花香
19、西面入口
20、东面入口
21、意澜桥
22、阳光草坡
23、木荷林
24、观演草坡
25、观景平台
26、攀爬场地
27、七步休息廊

休闲运动区
28、香泡林
29、水杉林
30、覆土管理室
31、运动场
32、极限运动场
33、背景绿林
34、五彩廊

重庆市九龙坡区
西彭组团J分区休闲公园

Jiulongpo District in Chongqing City
Leisure Park in Division J of Xipeng Group

项目类型： 休闲娱乐社区公园
项目面积： 6.6万平方米
设计单位： 重庆蓝调国际（绿茵景园集团公司）设计一所
项目地点： 重庆九龙坡区西彭
委托单位： 上海城开集团重庆德普置业有限公司
主　创： 任刚
团　队： 陈果、黄茜、赖斯、张俊、曹文婧
设计时间： 2010年

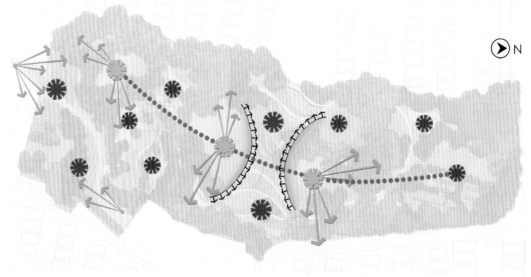

■ 后期视线分析　Visual Analysis in Later Stage

■ 缇香广场剖面　Profile Drawing of Tixiang Square

项目简介　Introduction to the Project

项目是2010年重庆蓝调城市景观规划设计公司为上海城开集团重庆德普置业有限公司提供的设计方案。公园位于重庆西部新城白市驿陶家片区，离大渡口较近。该区域正处于开发阶段，公园的建设将拉动区域的发展。公园周边用地规划是常青藤二期、三期高档别墅区，主要服务于周边住户，满足他们日常体育健身及休闲娱乐。在立足于周边高档社区环境中，设计师将项目定位于一个以森林生态、休闲运动、轻松趣味为主，服务于大众的社区公园。

The project is a design plan for Chongqing Depu Properties Limited Company of Shanghai Urban Development Group provided by Chongqing City Blue Tone Urban Landscape Planning and Design Company in 2010. The park is located in Taojia District in Baishiyi in the New City of the west of Chongqing City near Dadu Kou. This region is in a developing stage and the construction of the park will promote the development of the region. The land planning surrounding the park is for Ivy of the second stage and high-grade villas of the third stage. The plan serves for surrounding residents and satisfies their daily sports and entertainment. In the surrounding environment of a high-grade community, the designer orients the project as a community park which mainly contains the forest ecosystem, recreational sports and relaxing pleasure and serves the public.

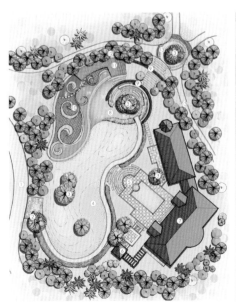

■ 缇香广场平面　Plan of Tixiang Square

■ 缇香广场效果图　Effect Drawing of Tixiang Square

城市公园绿地景观设计 | 重庆市九龙坡区西彭组团J分区休闲公园
URBAN PARK GREENLAND LANDSCAPE DESIGN

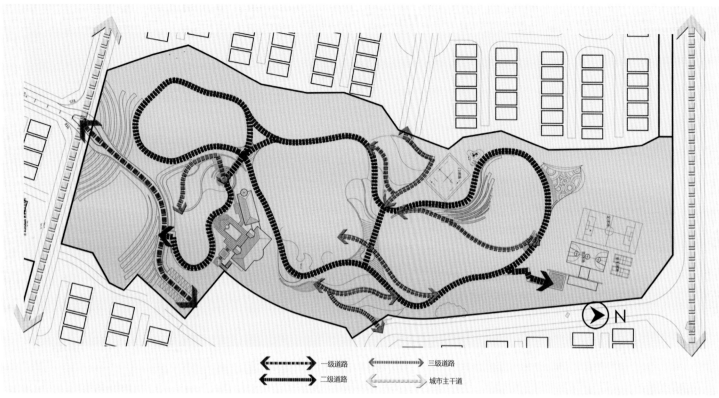

一级道路　　二级道路　　三级道路　　城市主干道

■ 交通组织　Traffic Organizing

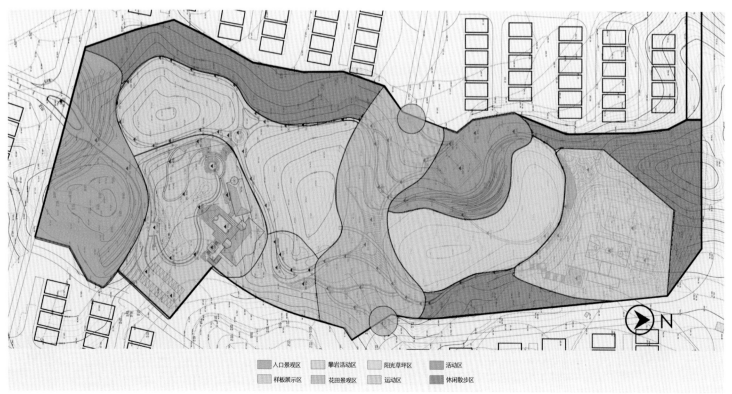

入口景观区　攀岩活动区　阳光草坪区　活动区
样板展示区　花田景观区　运动区　休闲散步区

■ 功能分析　Function Analysis

■ 花田效果图　Effect Drawing of Flower Field

■ 花田平面图　Plan of Flower Field

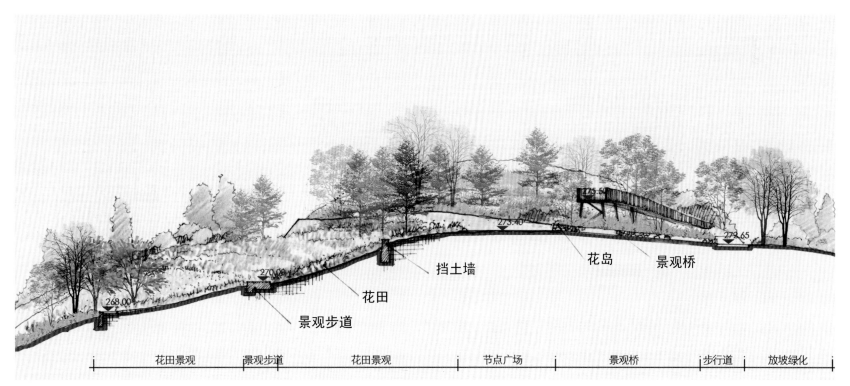

■ 花田剖面图　Profile Drawing of Flower Field

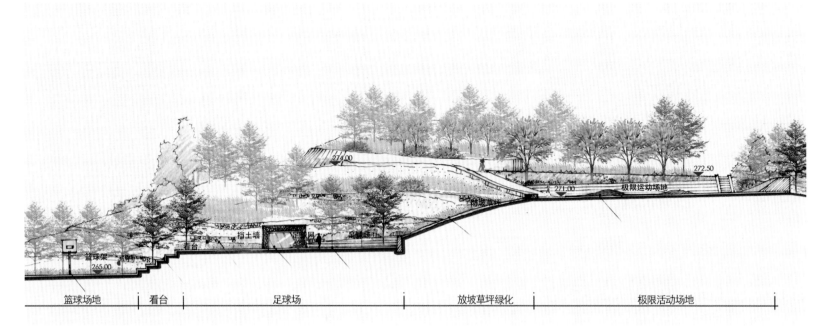

■ 极限运动场剖面　Profile Drawing of Extreme Sports Ground

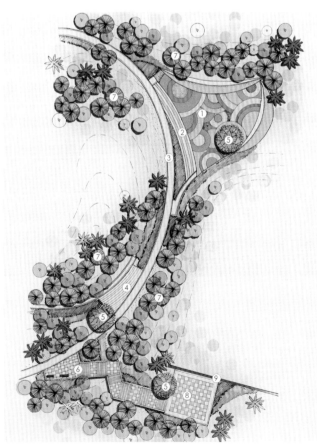

■ 极限运动场平面　Floor Plan of Extreme Sports Ground

公园入口通过宽阔的景观大道将公园藏于人的景观视线中，穿过密林，显现于眼帘的是大片花海与草坪构成的乡村意境。形状迥异的无边界游泳池和弧形景观墙成为公园景观中心的亮点。穿过景墙进入到了公园比较惬意的林荫道路空间。公园原始场地是属于低丘陵地形，且场地显南北走向，在狭长的丘陵环境中，设计师运用"项链"式设计概念串联起各个场所来取代传统的山脊动线，所形成的游憩环路增加了公园的游览路径。游憩路径通过动与静两种手法来组织整个路径的活动空间。

The entrance of the park hides the park in people's sight through the broad landscape avenue. Walking through the thick forest, people can see the village scenery with a wide range of flowers and grassland. The boundless swimming pools with a strange shape and the arc-shaped feature wall are the highlights of the landscape center of the park. Walking through the feature wall, people come to the cozy boulevard space of the park. The terrain of the original site of the park is a low hill with a north-south direction. In the environment of a long hill, the designer uses the design idea of "Necklace" to connect all the sites in place of the traditional ridge moving line. The surrounding road for sightseeing and resting formed by the design enlarges the tour road of the park. The road for sightseeing and resting organizes the activity space of the overall road through dynamic and static methods.

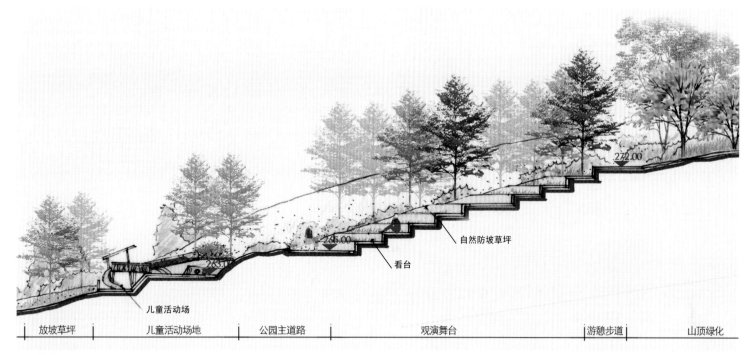

■ 观演舞台剖面　Profile Drawing of Theater Stage

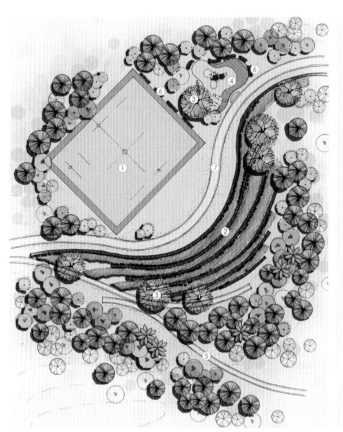

■ 观演舞台平面图　Floor Plan of Theater Stage

在"静"的空间中，鲜花烂漫的花海和惬意的林荫小道中，青石、土壤、凉椅无不让人忘却城市的喧嚣，共鸣于鸟语花香的自然场所。在"动"的空间中，儿童游乐场、篮球场、足球场、视线宽阔的观景平台给人宽广的胸怀去感受公园带来的乐趣。路径围绕的山顶以宽阔的大草坪形成的自由舒散空间表达对"自由、共享"意境的传达，创造出心意相随的共鸣空间。

In the "static" space, in the flourishing flowers and on the leisurely shady road the stones, soil and chairs can relieve people of the noise of the city and stays in a natural site together with singing birds and fragrant flowers. In the "dynamic" space, the playground for children, the basketball playground, the soccer field and the viewing platform with a wide sight enable people to enjoy the pleasure brought by the park with a broad mind. The mountain top surrounded by the road expresses the "free and sharing" artistic conception through the free and open space formed by the wide and large grassland. It also creates space where the mind and the environment are in response to each other.

旅游区景观设计
Tourism Area Landscape Design

可持续发展与因地制宜

旅游和房地产虽是完全不同的两个行业，但它们相互交叉、理念渗透，便形成了许多边缘性的全新综合结构。人人都渴望与青山碧水相伴，渴望呼吸到纯净的空气。较之一般的住宅，旅游地产的特点和优势在于它具有更好的自然景观、建筑景观，同时拥有完善的配套功能和极高的投资价值。对于旅游区设计，绿茵有着其独特的视角与触感，设计团队遵从"可持续发展的旅游产业链"这一理念，因地制宜，最大限度地保留景区自然风貌并与地产完美结合，努力让生态与人文同步发展。

■ 与原生态林地交融与一体的后花园
The back garden in combination with the original woodland

旅游区景观设计
Tourism Area Landscape Design

222	230	236
长白山国际旅游度假区南区 Apartment For Athletes In Changbai Mountain Internatinal Resort	保利·石象湖 Poly.stone Elephant Lake	中铁二局·花水湾度假小镇 China Railway Erju Co.Ltd.huashuiwan Resort Town

■ 干树万树红花开　Red flowers of thousands of trees in full bloom

风景印象派
中铁二局·花水湾度假小镇

Scenery Impressionism
China Railway Erju Co.Ltd — Huashuiwan Resort Town

项目开发商：成都中铁巴登巴登温泉投资开发有限公司
项目地址：成都市大邑县龙门山脉度假旅游带
总规划范围：10余平方公里，外围保护区600平方公里
项目类型：旅游度假地产
设计单位：成都绿茵景园　设计一所
方案主创：潘旭、张彬、陈辉
参与设计师：郭伟、孙斐、汪雅兰、李茂
施工图团队：曾繁利、潘迪、阙龙庆、粟凡粒、陈浩然
合作团队：新西兰LITTORALIS设计公司
设计时间：2009年

■ 川溪口鸟瞰　Aerial View Drawing of the Stream

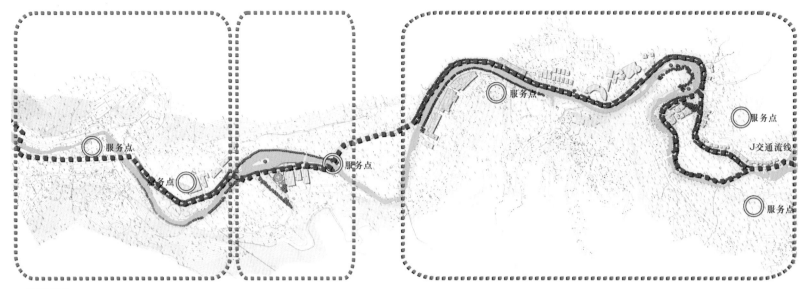

■ 区域分析图　Region Analysis Drawing

项目概况 Project Introduction

中国花水湾度假小镇地处成都市大邑县西南面，位于花水湾镇的核心区。总规划范围10余平方公里，外围保护区600平方公里，投资建设区域200公顷，预计总投资22亿元，建设周期6—8年。规划将其建设成为国家级旅游度假区，是成都市龙门山脉度假旅游带（东方阿尔卑斯山）的重点项目和示范项目。

本次设计内容为中国花水湾度假小镇溪流谷地景观规划。

China Huashuiwan Resort Town is in the core area of Huashuiwan Town located in the southwest of Dayi County in Chengdu. It has a planning area of 10 square kilometers with a surrounding protection area of 600 square kilometers. Its invested planning area is of 200 hectares and the planning investment amounts to 2.2 billion yuan. The construction takes 6-8 years. Constructing it into a national tourist resort is the key project and demonstration project of Chengdu Longmen Mountain resort zone (with Alps in the east).

This is a landscape planning design of streams and valleys in Huashuiwan Resort Town in China.

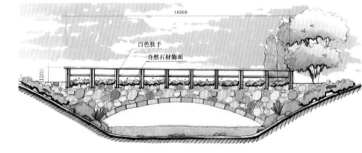

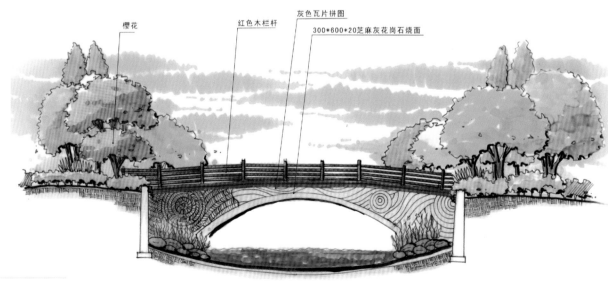

■ 《阿尔让特伊大桥》 Japanese Bridge

交通 Transportation

距成都市中心89公里，经成温邛高速公路、大双旅游快速通道仅70分钟车程。通过大新路可连接川西旅游环线S106。

自然资源 Natural Resources

典型的坡地山谷地形，水系丰富，森林覆盖率达81.7%。气候宜人，夏无酷暑，冬无严寒。拥有优质的医疗热矿泉水，以"人间瑶池"享誉全川。

社会经济 Social Economy

花水湾镇总人口8400人（2005年），以第三产业为重，是大邑县的旅游型城镇。

旅游发展目标 Tourism Development Goals

中铁二局以欧洲度假小镇为蓝本，结合旅游酒店、温泉、度假房地产、商业、运动、康疗、休闲娱乐等元素，构筑成结构完整的、先进的、可持续发展的旅游产业链，创造高品质"雪山——温泉"生态、健康、度假、商务四大旅游主题。规划常住人群容纳量为2万人，游客接待能力为100万人次/年。经成都市市场研究公司专题研究，预计度假小镇2009年达60万人次接待量。

Transportation

The resort is 89 kilometers away from the center of Chengdu City and it takes only 70 minutes to drive there through Cheng Wen Qiong Highway and Dashuang Tourism Fast Channel. It can be connected with Chuanxi Tourism Circle through Daxin Road.

Natural Resources

It has a typical terrain with slopes and valleys and abundant water systems. The forests cover 81.7% of the area. Besides, it has a pleasant climate that it is not extremely hot in summer or freezing cold in winter. It has hot mineral water of high quality and is famous in Sichuan as the Abode of Fairy Mother Goddess in Mortal World.

Social Economy

Huashuiwan Town has a population of 8400 (Before 2005) and emphasizes the third industry. It is a tourism town in Dayi County.

Tourism Development Goals

China Railway Erju Co.Ltd aims at constituting a comprehensive, advanced and sustainable developing torism industry line based on European resort towns, in combination with the elements of tourism hotels, hot spring, resort real estate, business, sports, health treating and recreation, etc. It also attempts to create four high-quality "Snow Mountain Hot Spring" theme tours of ecology, health, holiday and business. It is planned to contain twenty thousand permanent residents and one million person-trips each year.

规划设计策略 Planning Design Strategies

① 规划设计的目标 Planning Design Goals

高品质温泉养生胜地；
闲适旖旎度假小镇；
新锐户外运动大本营；
时尚先锋艺术殿堂。

打造贯穿中国花水湾度假小镇全域的溪流谷地景观带，形成集观赏性、功能性于一体的重要风景长廊。

High-quality hot spring health resort,
Leisurely and charming resort town,
Prominent outdoor sports base,
Fashionable pioneer art hall.

It aims at creating a landscape zone of lakes and valleys running through the whole China Huashuiwan Resort Town and forming an important landscape corridor integrated with entertainment and functions.

■ 桥上的风景 Landscape on the Bridge

② 规划设计的主题构思 Theme Conception of Planning Design

阿尔卑斯交响曲——植根花水湾场地本源之精神，撷取阿尔卑斯风景特色之精华，浓缩中西文化交融之精髓，以一曲澎湃磅礴的交响乐演绎溪流谷地传奇。

音乐是流动的，每个音符、每段乐句在时间上变幻跳跃，组成动听的乐曲。风景的美同样是随着时间的推移逐步展现在游人面前，每前行一步眼前的画面更替变化，所谓"移步易景"。

塑造风景空间和谱写音乐有异曲同工之妙。将一幅幅美丽的构图当作音符组成乐句进行创作，让溪流谷地的各景观节点如同流动的音乐一样，使整个观赏路线上有起始、展开、再现、高潮、尾声的序列演变，在风景空间的不同主题进行着独奏、变奏、重奏，乃至混合成山水立体交响史诗。

Alps Symphony is based on the original spirit of Huashuiwan site. A surging and majestic symphony which takes in the essence of Alps scenery character and concentrates the cream of culture exchange of China and western countries presents the legend of dreams and valleys.

Music is flowing. Every music note and phrase that changing and jumping as time goes by composes beautiful music. Similarly, the beauty of scenery is presented gradually over time. The picture changes as we step forward which is called scenery changes when steps moving.

Creating landscape space and composing music are rendered with the same skill. A beautiful construction design is carried out just like the notes are composed into music. The landscape joints of streams and valleys are like flowing music to make the whole tourist line have a sequence evolution of beginning, developing, reoccurring, high tide and ending. Different themes of landscape space are presenting different feelings just like hearing a solo, variation play, combination play or even a mixed threedimensional epic symphony of mountains and rivers.

③ 规划设计的风格定位 Style of the Planning Design

意境 Artistic Conception

虽以德国巴登巴登小镇为蓝本，但非追求一草一木的原样移植，设计师强调进行取样于斯的再创作，精雕细琢于景，融会贯通于情，力求创造出巴登巴登与花水湾共有的核心价值——闲适生活的意境美。

巴登巴登的久负盛名，缘由18世纪以来氤氲的温泉浴池，更缘由浸入每一个到访者骨子里的那份闲适。大文豪马克·吐温总结巴登巴登说："在这里，5分钟后你会忘掉自己，20分钟后你会忘掉世界。"夸张话语把闲适的感觉极致描述。

■ 桥与教堂相映成趣 The bridge and the church contrast finely with each other

■ 山水绿如蓝
The mountains and the rivers are very green, like the color of the sky.

■ 桥与河是花水湾绕不过的情结
The bridge and river are Huashuiwan's features that can not be neglected

闲适就是这样一种从容、自在、放松的感觉，什么都可以不去想，什么都不用去做。

找寻一个能让心灵放假的地方，那就到花水湾来吧。在这里，让你抛开尘世间的纷扰，让你放下快节奏生活的重负；在这里，连呼吸都感到心旷神怡，闲步观赏田园风景，品味美食佳肴，享受慵懒、惬意的生活体验。

景观印象 Landscape Impression

远眺晶莹的雪峰横卧在天际，川溪河蜿蜒淌过绵延起伏的山地，伴着清澈欢腾的溪水沿路徐行，两侧的坡地由上至下是浓密静寂的树林、繁花盛开的草甸、绿茵如云的河滩自然过渡伸向谷底，尽头高低错落的房舍依河而建，傲然卓立的教堂塔楼传来报时的钟点，桥梁、建筑的一砖一瓦都揉入欧洲小镇的浪漫情怀和艺术气息。

景观元素 Elements of the Landscape

小镇

浓缩了整个欧洲的灵魂，古朴典雅与闲情逸致来自于生活的平常和宁静，这种纯粹撩拨着渴望田园生活的都市人的心弦。

溪流 桥 石材 植被

Artistic Conception

Although the design is based on Baden Town in Germany, it does not copy the same style. The designer stressed the recreation sampling from the style. The landscape should be created with great care and the feelings should be achieved tin a comprehensive way. The designer attempts to crate a common core value of Baden and Huashuiwan, that is, the artistic beauty of a leisure life.

Baden has long enjoyed a good reputation for its hot spring pools with dense mist since the 18th century. It is also still more famous for its comfort and leisure which visitors immerse themselves in. The literary giant Mark Twain concluded with the saying that you will forget yourself in five minutes and forget the world in twenty minutes. The exaggerated saying has vividly described the feeling of leisure and comfort.

■ 别致的木亭为游人提供一处清凉之所
Exquisite wood pavilion offers tourists a place to enjoy the cool

■ 如画 Picturesque

It is such a feeling of leisure, comfort and relaxation with nothing think or to do.

If you are seeking for a place to relax yourselves, come to Huashuiwan. You can get away from the confusion of the world here and unload the burden of life. Even the breath can feel carefree and joyous here. Take a walk to appreciate the rural scenery and taste delicious food. You can enjoy a leisurely and comfortable life.

Landscape Impression

Overlooking the glittering snow mountains lying against the sky, we can see the rivers and lakes winding through the continuous and undulate mountains with the clean and joyous lakes flowing along. There are thick forests on the upper part of both sides of the slope and the flourish flowers on the lower part. At the foot of the slopes, the beach with many grass lands like clouds spread to the bottom of the valley in a natural way. At the end of the beach, the interlocking houses are built along the river and the church tower standing proudly spreads the striking of time. The bridge and all the buildings are con-structed with the romantic feeling and artistic flavor of European towns.

Elements of the Landscape

Town

The landscape concentrates the sole of the entire Europe. Simple and elegant leisure comes from the peace of life and the purity arouses urban people's desire for a rural life.

Stream, bridge, stone material, plants.

■ 下沉广场的风情之美 Beauty of Amorous Feelings of Sunken Plaza

详细设计 Design in Details

① 谷口门户区——田园抒情曲（明朗、清新）

这是华丽的第一乐章，一大片宽阔连绵、干净纯粹、起伏有致的草坪跃然眼中，满目青翠欲滴，奏出河滩与溪流交融、山水与小筑相依的旋律。游人如织，顾盼相望，随着视线越来越舒展，曲调也越来越优美。教堂钟声在恬静、平和、自然的田园抒情中达到第一个高潮。

风景幽静的一隅，隐约可见尖尖的钟塔，那是在欧洲小镇里不可缺少的一部分——教堂，但凡有了它的存在，小镇才是完整的。晚钟悠扬，音色美妙，在恬雅闲适、诗意朦胧之间，顿悟精神境界的超然。

② 静水焦点区——水的变奏曲（典雅、变幻）

平矶曲岸、小岛长堤把水面划分为既隔又连、层次丰富的空间，静水流瀑在这里变换着曲调。平静的池水如一面明镜，岛树倒映，光影交叠，虚实变幻。漫过水坝，水激砥石，随着水量大小变化或铿锵有力地轰鸣，或潺潺弱弱地低吟。

③ 峡谷区——集市小步舞曲（活泼、轻快）

道路一侧山崖陡立，另一侧悬在溪流之上。集市隐藏在对岸的绿树浓荫之中，巴洛克风格的建筑围合着用鹅卵石铺砌的小广场，泛着青苔的喷水池汩汩泉涌。狭窄的小街两旁，人字形的小屋涂上鲜艳的色彩，窗台是美丽的花篮。油漆斑驳的小酒吧、小巧精致的咖啡馆以及手工小作坊令人忆起狄更斯笔下所描绘的中世纪的生活画面……

④ 谷中开发区——白鹭回旋曲（恬静、悠扬）

漫山松林苍翠，一丛丛迎春、一簇簇野蔷薇灿烂绽放，柔和了山的线条，软化了石墙的坚硬。走进寂静中的山路，树影间洒下点点光斑，溪边水草透出晶莹色泽，一行白鹭久久回旋……

■ 与景区植被呼应的外围标识墙　Outlying Sign Wall in Combination with Scenic Vegetation

第二次生命，风景区景观再营造
保利·石象湖
The Second Life, Recreation of the Landscape in Region

开发商：保利(成都)石象湖旅游发展有限公司
地点：成雅高速公路66公路出口、国家级生态示范区成都市蒲江县石象湖景区内
设计面积：30000平米
概念设计：美国SWA景观设计公司
景观设计：成都绿茵景园　设计二所
主创：阿笠
方案设计团队：张洪源、刘鸠鸠
施工图团队：董玲、袁怡、潘宇、段倩
设计时间：2009年
完成阶段：施工完成

■ 景区入口门卫房设计图　Design Plan of Guard Room at the Entrance of Scenic Region

设计特色 Design Features

本项目地块位于石象湖景区内，自然地被条件较好，林地保护完善，有利于场地景观带的营造；主入口通过一面长200米、高度为6米的景观墙，将高速路的干扰挡在园外。而这面景观墙虽然有6米高，但通过采用镂空、打断处理，并配合土壤造坡，使其融入背景林里，所以并不会感觉压抑。道路线条流畅，并在道路中间设置绿岛，形成林荫的道路系统。而会所周边增加自然湖泊，流畅的驳岸线条配以自然河石，岸边的水杉倒影在明净的湖水里，这一切都净化着人们的心灵。

The project is located in internal Stone Elephant Scenic Spot. It has a good natural ground condition and well protected woodlands which are beneficial for the creation of site landscape. There is a landscape wall with a length of 200 meters and a breadth of 6 meters in the main entrance. The wall keeps the disturbance of the high way off the garden. Although the wall is 6 meters high, it will be in harmony with the background wall by dealing ways of hollowing and breaking and by combining it with the earth slopes. So the wall will not bring depressing feelings. The road line is smooth and the green island is set in the middle of the road to form a shadowy road system. Besides, the natural lakes are added surrounding the club and smooth bank line is decorated with natural lake stones. The sequoias along the bank are mirrored on the lake and purify and brighten people's mind.

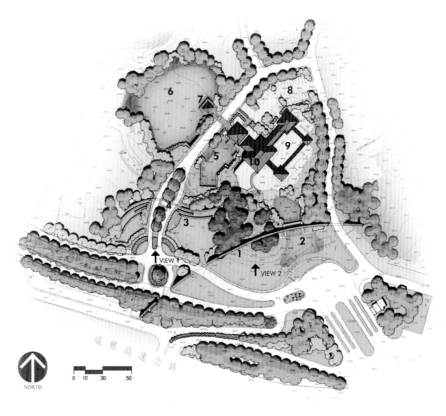

■ 景区入口景观改造设计平面　Design Plan of Landscape Transformation at the Entrance of Scenic Region

■ 外围水景与标识墙生态的互融
The Combination of Outlying Water Landscape with Sign Wall Ecology

■ 与景区原生林地叠加在一起的人工湖
Artificial Lake Stacked with Ecological Forestland in Scenic Region

■ 景区入口门卫房实景
Actual Scene of Guard Room at the Entrance of Scenic Region

■ 景区入口水景标志墙
Water Landscape Sign Wall at the Entrance of Scenic Region

■ 景区交通入口立面
Elevation Plan of Traffic Entrance in Scenic Region

■ 景区标识景墙体立面效果
Elevation Effect of Sign Wall in Scenic Region

■ 会所外的草地小景
The Little Scenic Grassland Outside the Club

■ 会所后花园水景台地
Water Landscape Platform in the Back Garden of Club

■ 会所前门水景池
Water Landscape Pool in front of the Club

■ 从会所花园远眺景区原生林地
From the Club Garden, looking at the Ecological Forestland in Scenic Region

■ 会所边原生的林地植被
Ecological Forestland Vegetation besides the Club

■ 坐落于山丘之上犹如庄园般的会所
The Club Located on the Mountain like a Manor

■ 会所入口令人回味的水景
Water Landscape with a Long and Sweet Aftertaste at the Entrance of the Club

■ 会所周边大量的花卉
Abundant Flowers besides the Club

旅游区景观设计
TOURISM AREA LANDSCAPE DESIGN

长白山国际旅游度假区南区

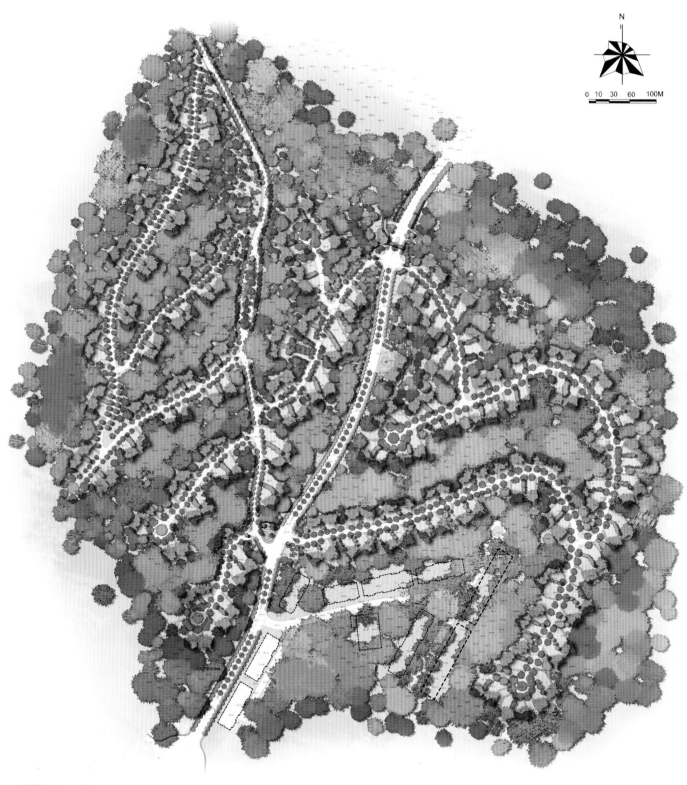

■ 总平面　General Plan

长白山国际旅游度假区南区
运动员公寓、酒店式公寓

Apartment for Athletes in Changbai
Mountain International Resort, a Hotel-styled Apartment

项目名称： 长白山国际旅游度假区南区运动员公寓、酒店式公寓
项目类型： 旅游度假式高级别墅群
项目面积： 景观设计总面积是37.1万平米
设计单位： 重庆蓝调国际（绿茵景园集团）设计三所
项目地点： 吉林省白山市
委托单位： 长白山国际旅游度假区开发有限公司（万达集团）
主　创： 李锦香
团　队： 肖勇、张骊瑜、熊扬眉、王有龙、左春丽
设计时间： 2010年8月

■ 门楼　Arch over a Gateway

项目简介 Introduction to the Project

　　长白山国际旅游度假区南区运动员公寓、酒店式公寓是蓝调公司在2010年国际竞标中获胜的一大力作。长白山国际旅游度假区是中国最大旅游投资项目，由大连万达集团、中国泛海集团、内蒙古亿利资源集团、辽宁一方集团等四家中国顶级民营企业联合投资200亿元打造。长白山国际旅游度假区位于白山市抚松县松江河地区，距抚松县城23公里，距长白山机场约10公里，距长白山天池只有68公里，长白山中国一侧的最高峰——白云峰就坐落在本区内。长白山国际旅游度假区分南北两个区，北区以旅游服务为主，南区以体育运动休闲为主，是2012年全国冬季运动会分赛场之一，在南北两个区之间还夹着松江河镇和东岗镇。南区规划用地面积13.3平方公里，包括滑雪场、商业街用地、星级酒店区、娱雪区、雪上两项小球训练中心、运动员公寓、酒店式公寓及大型人工湖区等等。本次设计位于南区，为生态型、运动型超低密度的北美风情高级别墅社区。

■ 北入口一立面　Elevation Drawing of North Entrance One

Introduction to the Project

Apartment for Athletes in Changbai Mountain International Resort, a hotelstyled Apartment was a project which Blue Tone Company won in the International Bid in 2010. Changbai Mountain International Resort is the largest tourism investment project in China, created with the joint investment of 20 billion yuan by four top private enterprises in China which are Dalian Wanda Group, China Ocean Wide Group, Elion Resources Group in Inner Mongolia and Liaoning Yifang Group. Changbai Mountain International Resort is located in Songjianghe Region in Fusong County in Baishan City. The resort is 23 kilometers far away from Fusong County, about 10 kilometers far away from Baishan Airport and only 68 kilometers far away from the Heaven Pool in Changbai Mountain. Baiyun Peak, the highest peak in Changbai Mountain on the side of China, is located in this region. There are two divisions in Changbai Mountain International Resort, the north division and south division. There is mainly traveling service in the north division and sports and leisure in the south division. The south division is one of the playing courts of National Winter Games in 2012. Songjianghe Town and Donggang Town are located between the two divisions. There is a planning area of 13.3 square kilometers in the south division, including a ski resort, land for the commercial street, star-rated hotels region, entertainment, playing area with snow, the training center of two balls in the snow, the apartment for athletes, the hotel-styled apartment and a large artificial lake. The project is located in the south division and is an ecological, sporting and low-density high-rate villa community in a North American style.

设计构思 Design Idea

本案是长白山下的北美风情别墅，设计师充分结合了东北地区的地域特征，同时把北美自然风情的特点及度假休闲的生活方式充分表达。景观在细节的设计上顺应地形，植被选用当地树种，场地布置尽可能的南朝向，避免了冬季的严寒。景观的主题更多的体现北美的度假休闲生活。设计将北美最具代表性的加拿大洛矶山脉下的冰原大道及该区域的几处国家公园命名别墅区内的花园组团。由"冰原大道"划分为"杰士伯花园"与"斑芙花园"两大主题花园，体现森林深处的度假生活方式。在长白山森林深处，坐落着自然的北美林中的木屋，金色的松桦林海下，有着闲适的高尔夫生活以及动感十足的滑雪体验。旨在营造一种全天候的度假生活，景观细节与地形有机契合，人与自然零距离接触。居于松桦的金色背景及长白山林海雪原的地域特征，创造出独特及令人难忘的景观体验。

设计目标：打造一个北美自然风情的生态型经典高档别墅区。

In the design of the subject under the Changbai Mountain in a North American style, the designer integrates closely the features in the Northeast area. Meanwhile, the designer sufficiently expresses the features of North American style and the leisure lifestyle for holiday. In the design of landscape details, local trees are selected as a vegetation cover and the site is mostly set in a north orientation to avoid cold in winter. The landscape theme more the leisure life for holiday in North America. The d names the garden groups in the villa region with the representative Icefield Parkway under Canadian Rockies and national parks in this area. The "Icefield Parkway" is divid two theme parks, "Jesper Park" and "Banff Park", pre the lifestyle for holiday in the deep forest. — In the deep f Changbai Mountain, there is natural cabins in North African and in the golden birches, you can enjoy a leisure golf l dynamic skiing experience. The design aims to create a holi for all day long, make an organic combination of landscape and the terrain and enable people to contact closely with natu landscape experience is unique and unforgettable with the lif golden background created by birches and the re characteristics of forests and snow land in Changbai Mountain

Design Object: to create a typical ecological high-gra region in a North American style.

■ 班夫花园门岗立面 Elevation Drawing of Guard Room of Banff P

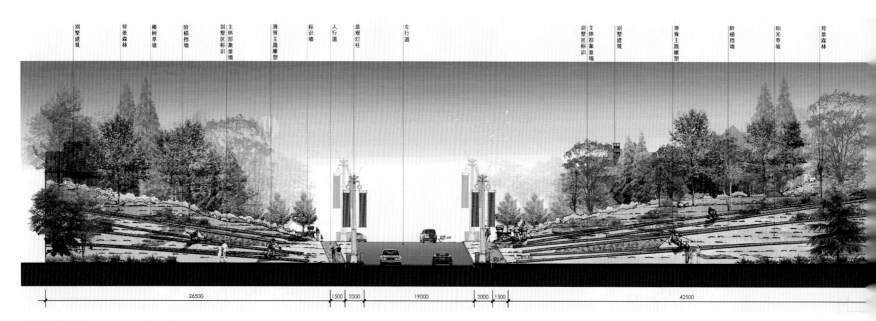

■ 北入口立面 Elevation Drawing of North Entrance

→ 总体设计 Overall Design

1、总体风格为北美自然风格。

2、总体竖向设计：在道路及入户标高的基础上保证大地形的完整，减少挖填方，力求营造自然的山林效果，同时为业主提供更好的观景视线。

3、总体种植设计：种植设计将考虑到长白山的气候和土壤特征，在满足整体风格和效果的前提下尽量选择乡土树种。总体上对别墅进行组团式划分，根据不同地块的展示期主题花园选择其展示期间效果最佳树种作为主题树，同时通过统一的行道树、背景树和景观骨干树等将各个区域有机地协调起来。

1. The overall style is a North American style.
2. The overall vertical design: Based on the residential elevation and roads, the design should ensure the completeness of the overall terrain. Digging and filling should be reduced to create a natural effect of forestland and at the same time provide the residents with a better viewing sight.
3. The overall planting design: The planting design takes the climate and soil features of Changbai Mountain into consideration. It also mostly chooses local trees on the premise of satisfying the integral style and effect. In the general, the villas are divided into groups of land. According to different exhibition periods of different plots of land, the theme garden should choose the most appropriate species of trees as theme trees which have the best effect in the exhibition period. Meanwhile, all the regions should be organically coordinated through unified roadside trees, background trees and main trees of landscape.

■ 北入口二放大平面 Enlarged Ichnography of North Entrance Two

→ 具体设计 Details of the Design

① 景观大道（冰原大道）Landscape Avenue

该大道是进入别墅区的主要交通道，是城市区域进入别墅区的第一个展示面，形成对别墅的第一感受。景观在入口区域以门楼的形式形成视觉亮点。体现出高级精品别墅的高贵气质。再以两排高大行道树及中央绿化分隔带加强森林的林荫效果；进入别墅区内在服务中心及交通转换节点通过塔楼、跌水景观、北美拱廊形成视觉高潮。

1. Landscape Avenue (Icefield Parkway): The Avenue is the main traffic road for people to enter the villa region. It is also the first exhibition of the villa and gives people the first impression for the villa when entering the villa region from the city. There is a highlight of the landscape in the entrance area in the form of an arch over a gateway. The avenue presents the nobility of high-grade villas. Then, the shady effect of the forest is strengthened by tall trees on both sides of the avenue and central green dividing strip. When people enter the villa region, they will have a visual climax through towers, landscape of water drop and North American arcades in the service center and transfer joints of traffic.

■ 塔楼 Tower

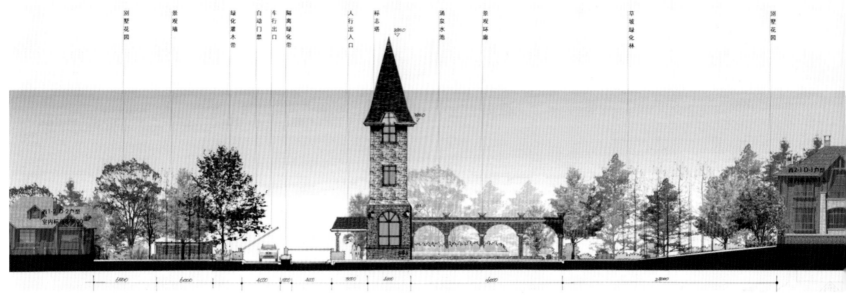

■ 南入口剖面　Profile Drawing of South Entrance

② 康纳特大道 Connaught Avenue

　　通过主入口景观大道大体量的门楼建筑及装饰大楼，进入第二级的景区内道路则要为业主创造温馨的归家氛围，采用精致的点景大树及景观大乔木，以营造尊贵、自然的北美生活感。

③ 组团花园 Groups of gardens

　　采用自然的种植手法，通过有层次的种植呼应背景森林的效果，尽量减少不必要的土方填量和挡土墙。在入户车道的两侧设置景观微地形，创造自然起伏的地形变化，同时也可作为隔离之用，为临路的别墅创造更多的私密性，并减弱交通噪音的干扰。采用自然材料(石、木)增加艺术气息，形成安静的自然与人工相结合的绿色休息空间。在临近高尔夫的区域采用自然有间隙的种植手法，使景观视线通透，借景高尔夫球场的景观。 通过地形的塑造创造丰富多变的视觉感受；公共绿化区域作为关键节点设置景观兴奋点，运用开敞的林间草坪、叠落式植草露台等形式创造若干不同的特色空间，用于室外运动及集会、室外艺术品展示、时令花卉展示、风景眺望等功能；在从社区内部进入公共绿地（小公园）的出入口处设标志柱及可控制的装饰门。

2. Connaught Avenue: People enter the secondary scenic region through large gatehouses and decorative buildings along the entrance landscape avenue. The roads in the secondary scenic region provide owners with the comfortable returning atmosphere. The exquisite featured trees and tall landscape trees create a noble and natural sense of living in North America.

3. Groups of gardens: A natural planting method is taken and the planting with different levels is in coordination with the effect of background forest. Unnecessary use of earth filling and retaining walls should be reduced. The setting of landscape tiny terrain on both sides of entrance driveway creates naturally undulating terrain changes and at the same time, it serves as isolation to create more private space for the villas along the road and weaken the disturbance of traffic noise. The use of natural material (stone, wood) adds an artistic flavor to create quiet green resting space with a combination of nature and artificiality. In the region near golf court, the natural planting method with gaps makes the landscape sight without blocking with the help of the golf court landscape. The shaping of the terrain creates various visual feelings. The lot of excitement is set with the public green area as a key point. Besides, the forms of open grassland in the forest and an interlocking terrace with grass create various featured space to serve for outdoor sports and meetings, outdoor art exhibition, seasonal flower show and sight overlooking. At the entrance of the public grassland (a small park) in the internal community, there is a mark pole and controllable decorative door.

■ 廊琴花园效果图　Effect Drawing of Langqin Park

■ "端头路"式私家花园入户方式效果图　Effect Drawing of the Residence with a Private Park in a "Leading Road" Style

④ 别墅私家花园　Private gardens in the villa

在主要功能区的朝向上，严格尊重地域气候条件，使每户休闲场所尽量南朝向，兼顾到主要功能空间与私家山林的联系，考虑到环境与建筑的结合，入户形式做到多样化趣味性，装饰构筑物、小品、雕塑、灯具、楼栋号、垃圾桶、导示牌、信报箱、灯具等设计在细节处体现出北美风情。

总之，我们希望创造一个有无限自由空间想象和丰富度假体验并具有北美风情的高档别墅住宅区。

4. Private gardens in the villa: The orientation of main functional regions strictly follows the climate in the region and tries to make the leisure place in every house face south. The relation of main functional space and private mountain forest has been taken into consideration. As for the combination of environment and the architecture, the form of the residence is diverse and pleasant. The design details in decorative structures, scenes, sculpture, lamps and lanterns, floor numbers, trash cans, guidance plates and letters and paper boxes present the flavor of North America.

In conclusion, we hope to create a high-grade villa community with free space and imagination and rich holiday experience in a North American style.

■ "端头路"式私家花园入户方式效果图-雪景
Effect Drawing of the Residence with a Private Park in a "Leading Road" Style

特别专题
— 酒店景观设计
Special Column
Landscape Design Hotels

酒店（HOTEL）一词来源于法语，当时的意思是贵族在乡间招待贵宾的别墅。时至今日，酒店市场定位已不单单是招待贵宾，其市场定位已更加专业化、多样化。

现代酒店设计有外向性及与环境的一体化的特点。更加注重自然环境、人文环境与地域特色。经营与消费活动的室外化，酒店的主题化和品牌化，均要求设计师自觉地去尊重与欣赏环境，将酒店建筑本身的定位、体量以及空间与功能组织，做恰当的处理与把握。

现代景观设计在酒店的策划与设计建造过程及将来的运营中，均扮演着重要的角色。充分实现酒店建筑与周边环境的协调与统一，为将为酒店长期良好的运营创造有利的条件。这是设计师在酒店设计中需着重把握的。

青城（豪生）国际酒店景观设计
Howard Johnson Landscape Design

| 246 | 254 | 260 | 270 | 290 | 310 | 322 |

- 解读青城 Understanding Qingcheng Moutain
- 酒店设计概念 Hotel Design Conception
- 酒店入口景观 Entrance Landscape
- 酒店标准景观 The Hotel Standard
- 温源谷 Wenyuan Valley
- 道缘休闲岛 Daoyuan Leisure Island
- 高尔夫挥杆练习场 Golf Driving Range

The Story of the Mountain
青山之幸语

文/阿笠

每座山都有自己独特的语言，山之所以为山，因为它不仅仅是堆土坡，它是自然形成的群体，而群体的布局、组合、形态都有自己地形地貌的特色和文化。每座山都有自己风雨沧桑的记忆，和能够激起这种记忆的一些遗存的风景、文化，也就是说每座山有自己的语言。

走在巴黎，绝对不会以为走在伦敦，在纽约你绝对不会误以为是在洛杉矶，北京和上海不同，重庆和成都不同，历来如此。这种不同是历史文化和人文的沉淀，是这座山世世代代传下来的记忆的结晶。可惜的是，在崇尚外国的月亮比中国圆的今天，唯独缺少的是自己语言的传承。现在好多山自己的语言渐渐消亡，没有了自己独特语言的山是悲哀的。

山毁了，再多的钱也造不回来了，千百年历史沉淀出来的山之语没了，谁还能再创造出那一个个曾经灿烂的文化来？这些年来，国内大型开发公司希望能将这种山的文化传承下去，但是，这种商业行为成功与否还有待时间的检验。而青城国际酒店这个项目之所以能坚持下来，是在很多山之语将要消失殆尽的时候，凭借着那么一丝丝没有断的气息，奋力的向青城山的山韵突围。

在历史名山中开发，如何处理好保护遗产和开发的关系，对于任何一个国家的建筑师、景观设计师、规划人员来说，都不是一件容易的事情。除了对历史要有更加深刻的认识之外还要不断了解一些相关的资料，同时也要注意和其他专家、学者的交流，特别是那些对历史文化保护有经验的学者专家。他们会给我们很多新的思维、新的咨讯，自己在其中也有很大的提升。当然，对于青城山的文化和历史，我有一种敬畏感，因为实在太博大精深了，在设计时，尽量多综合各方面专家的参考资料。

在接手这个项目之时，我作为这个项目负责人深感其中的挑战，不仅是对自己技艺使用上的考验，也是对如此重要的文化的保护与致敬。虽然过程中也会挖掘其历史文化，展现其重要的道家精髓，但是我们感到这个项目更多的是需要将青城山的历史文化和道缘哲理结合在一起，在展示不同内涵的同时也要启迪到来的游人们与山的共鸣。

然而就在连续的设计、交流讨论中，心中逐渐的产生了一种疑惑：为什么而设计？

我们一直非常重视"设计"这个词。现在，"设计"已经成为一个在全世界范围内频繁出现的词汇，已经融入我们生活的点点滴滴，随处可见的设计充斥着我们的生活。设计到底是为了什么呢？

著名设计师阿部雅世曾说过："我们应该和那些有着同样危机感的人携手并肩，逆流而上，去找寻那些曾经被自己置之不理的理想观念。在那些设想完全埋没之前，用双手把它们挖掘出来，让自己敏感起来，去体会封存于其中的哲学。而且要重新构建有框架的日常生活，并将它代代相传下去。"

这种疑惑一直持续到2008年5月12日汶川大地震。当时，正好在青城山工地上，经历那一场本世纪以来最大的灾难，地动山摇，昏天黑地，见识到了大自然的威力，人类渺小的就好像蚂蚁，只能对突然爆发的灾难逆来顺受，毫无反抗之力，就在几秒中，从山顶到山底建筑轰然倒塌，居民流离失所。然而在接下来的救灾过程中让我感受到了另一种不曾感到的力量，见识到了一种人性互助的光辉，以及与自然仍然和谐共存的生活方式。抗震期间，身穿土布衣服、脚步泥泞的庄家汉，背个大背篓忙碌往返于受灾的家和临时居住点之间，以朴实的行动，将掩埋在大山中的家通过背上的背篓与外界联系。

是啊，这种朴实的行为是我们苦苦找寻的一种原始思想，一种无论何时何地都要如大山一样坚强屹立的坚持。而这种思想已经融入当地老百姓的骨髓里，反映到他们的日常生活、行为中，只是我们没能静下心来好好感受。而这种自强不息的感动，和全国各地自发的捐款，让全中国上下人们心系一线，感同身受。让我们感动、感悟、感激、乐观面对。突然领悟到现在这个工程，不仅是个工程而已，而是一个非常重要的记录这个事件的承载物，同时也是这段历史的重要记忆片段。而我们要做的就是将现在这种感受和感动融入到设计，符合这个含义。因为我们肩负一种历史责任，除了修复环境外，更多是带给那些受伤人们抗灾的信心！那么工作也就不是平常的工作，而是"精神"的设计！

西方人讲究的是对抗自然，东方人是共融与传承，除了把上千年的文化和环境修复成以前的光景，同时也要连接崭新的未来。我们同各类专家一起走访、研究和探讨。联想到长城是与山脊结合得很好的世界遗产，都江堰的鱼嘴是和水结合的很好的水利工程，在这次大地震中祖先所遗留下的许多古老工程都安然无恙，它们千年来经历的灾害岂止这一次，但与自然和谐的共存使它们至今屹立，并已然成为了自然中的一部分。回到现在来反思，不应放弃和丢弃老祖宗与自然和谐共生的理念，不是对立，而是如何疏理，引导它。所以我们的设计目标就是如何引导项目与环境更加的和谐共生，顺接千年文化，并一直传承下去，同时应该为可持续发展预留出人性化的生活和开发空间。从而更加尊敬这座山的意识形态来讲，不要妄想改变山，而是要传承它。灾后更多的思考能让人感慨、感悟和坚强。我们没有遗忘他们，是山之幸福，更是人之幸福，这就是设计觉醒的力量。

之前一年曾去日本考察，最让我震撼的是日本阪神殿，在其鼎盛时期也曾发生地震，一夜之间全毁，但是在之后的几十年间却能做到一砖一瓦全部恢复并和从前一样繁华，甚至更加牢固更加美丽，让现在的人感受到另一种感受。在灾后青城山快铁开通后，青城山作为成都人的后花园，其地位不但没有减弱，同时让更多人感受到这里的坚强，青城山在灾后与人们融合得更加亲密和谐。

回到项目设计的初衷，景观设计的力量是推动社会前进的力量和历史使命感。它推动着我们前行，让我们遇到困难的时候，支撑着我们鼓起勇气振作起来，克服过程中的种种困难。

至今，项目阶段性完成了，回首之前这两年的设计施工进程，与大家分享其中的滋味。本书中我将把这两年的设计历程一一展现给大家。

The Story of the Mountain

A Li

Every mountain has its unique language. The reason why a mountain stands as a mountain, is by no means that it is a mass of slopes and stones, but it is a naturally formed collection. The layout, arrangement and shape of the naturally formed collection are with particular land characteristics and culture. Every mountain has its memory of vicissitudes and is with the landscape and culture that can wake the memory and those are what we call as its language.

Just as when you walk in Paris, this is no way you will think you are in London, and if you are in New York, you won't believe it is Los Angeles. Beijing differs from Shanghai and Chongqing and Chengdu are not the same. This difference is from the history and culture sedimentation, just as a mountain's accumulation of its memories for centuries. It is a pity that in the modern times when people tend to idolizes foreign civilization and culture, what we lack is the carrying on our own language. Nowadays, as the agonizing case, a considerable number of mountain languages are fading. A mountain without its unique language is just pathetic.

Once the mountain is ruined, there is no way to reconstruct it at any cost. If the mountain language for centuries is gone, there is no possibility that anyone can create the once marvelous culture. For years, the largescale real estate companies cherish the hope that they will carry forward the culture of mountains, yet their commercial conducts are still waiting for the examination of time. The reason why the project of Qingcheng International Hotel went on and was succeeded, is that it stuck to seeking the traditional culture of Qingcheng mountain and thus presented its the language again.

In the development of historical mountains, it is never an easy task to strike the balance between heritage protection and development. Besides the profound knowledge about history and relevant information, communication with some other experienced experts and scholars in historical and cultural protection are significant, for we will be presented with new ways of thinking and thus gain much. Of course, confronted with the history and culture of Qingcheng mountain, I am filled with awe for its profoundness and referred myself to the relevant files and consulted experts of all fields while doing the design.

When I undertook the project, I felt deeply inside the challenge as a project director. It was not only a test of my techniques and skills, but also a protection and solute to such an important culture. Though in the course we would dig more about its history and culture, and display the significant essence of Daoism, we were more aware that the project was supposed to combine the Qingcheng culture and history with philosophy of Daoism and enlighten the coming tourists while showing its unique connotation.

But in the continual designing, discussion and communication, a question gradually occurred to me that what we were design for.

We used to pay much attention to the word design, and it is a frequent used word all around the world and is part of our lives. Then, what, on earth, is design for?

A celebrated designer once said, "We are supposed to be hand in hand with those who are with the same sense of danger and go upstream to find the ideal concepts once discarded by us. Before they are completely buried, we can dig them out with hands and wake our sensitiveness and taste the philosophy in them. What's more, we need to reconstruct the daily life with frames and get them to the later generations."

The question lasted to the Wenchuan Earthquake on May, 12th, 2008. I was on the project site to witness the most shocking catastrophe since the beginning of this century. I saw the nature's disa-strous power and human's fragileness and in several seconds the houses on the mountain just fell apart, and people were homeless. However, in the later rescue, I felt a strength that I had never tasted, and I saw a brightness of people's mutual help and a way of living of harmonious co-existence with the nature just as before. The farmers with homemade clothes and muddy feet, should-ered pack baskets, came and went between their buried homes and temporary inha-bitancy and made a connection with their simple actions.

That is it. The simple actions are the original spirit we pursue, a persistency as a mountain wh-atever happens. And the spirit had melt into the minds of local people and was in their daily life which we failed to feel with a placid heart. The uncasing striving and donation from all corners of the country made all the people fight and feel as one man and face the reality optimistically. It suddenly dawned on us that this was more than a project, but an important record for the event and history. What we needed to do was to combine the movement and feeling into the design, for we shouldered a historical responsib-ility to bring the injured people confidence besides repairing the environment. In this way, our work was no more a normal one, but a design of spirit.

Western people advocate to fight and change the nature, while easterners adore co-existence and to carry forward heritage. In addition to repairing the traditional culture and environment to what they were, we were also supposed to connect the project to the brand new future. Together with experts of all fields, we visited, studied and discussed. The great wall with the mountain ridge is a great world heritage, and the fish mouth of the Dujiangyan project with the stream is a fabulous irrigation works. A mass of antique projects from our ancestors were still sound and their experiences of calamities were far more than just this time. It is the harmonious co-existence with the nature that ensures they still stand and become part of the nature. Our reflection was that to give up the harmony co-exis tence concept of our ancestors was not wise and it was advisable to carry on the concept and leave the room for sustainable development. From the point of view of respecting the mountain, to carry on the heritage is wiser than to change it, and in addition, combining the movement of being self-striving in the rescue, all of these are in fact, the strength of the new reflection for the design.

In my visit to Japan one year ago, HANSHIN temple was most impressive. It was completely ruined in an earthquake in its prosperous times. What affecting was that it stood again in the later years with even firmer construction and more ma-gnificent look. After the Qingcheng express was in operation, Qingcheng mountain, as the back-yard of Chengdu people, stuck us with an image of self-striving and was closer to people in mind.

Back to the initial purpose of project design, it is the power of landscape design that propels the development of society and strengthens sense of historical responsibility. It cheered us up while encountering toughness and thus we co-nquered difficulties.

Up to now, the project has come to a stage end. And looking back the 2 year designing course, with this book, I am about to share and present our experience.

青城山

便利的交通条件
CONVENIENT TRAFFIC CONDITIONS

青城山东距省会成都市区50公里，是离成都市区最近的世界级旅游度假区，距青城山仅58公里的双流国际机场和已经建成的成灌高速、成青快速通道等交通网络将为规划区的发展带来极大的便利。以规划区为圆心6小时的车程距离可以覆盖四川境内所有世界级旅游风景区。

项目地理位置情况及历史文化背景
PROJECT LOCATION CONDITIONS HISTORICAL AND CULTURAL BACKGROUND

青城山是中国首批公布的风景名胜区之一，距都江堰市区16公里。青城山靠岷山雪岭，面向川西平原，群峰环绕，状若城廓；林深树密，四季常绿；丹梯千级，曲径通幽。公元143年，『天师』张陵来到青城山，选中青城山的深幽涵碧，结茅传道，被道教列为『第五洞天』，至今完好地保存有数十座道教宫观，珍藏着大量古迹文物和近代名家手迹。青城山是一座纵横千百年的活的道教『博物馆』。青城山后山景区更有大蜀王遗迹、遍布山中的飞瀑渊潭、曲桥栈道，颇具原始野趣。青城山的人文景观、自然风光交相辉映，故有『青城天下幽』之美誉。主要景点：建福宫、天师洞、掷笔槽、访宁桥、祖师殿、上沩宫、老君阁、玉清宫、月城湖、金鞭岩、水晶深洞、泰安寺、金娃娃沱、三潭雾泉、龙隐峡栈道、又一村、白云万佛洞、通天洞、白云古寺、白云古寨、双泉水帘、翠映湖、百丈长桥等。青城山的保护和繁荣同中国道教的在这里的兴起和发展有着密切的联系。

独有的自然资源
UNIQUE NATURAL RESOURCES

青城山以其自然景观和人文景观的幽古清雅博得『青城天下幽』的美誉，与剑门之险，峨眉之秀，夔门之雄齐名。景区面积200平方公里，山上古木参天，四季常青，拥有原生态植被群落，空气质量和水质常年保持国家一级水平，空气中负氧离子含量高达91%，是少有的天然氧吧。

千年人文景观
MILLENNIUM HUMANISM LANDSCAPE

拜水都江堰问道青城山

古老的都江堰水利工程被誉为『世界水利文化的鼻祖』，有『青城天下幽』之称的青城山，是中国道教发祥地之一；青城山都江堰已成功列入《世界遗产名录》，成为世界著名的旅游胜地。

青城（豪生）国际酒店景观设计
HOWARD JOHNSON LANDSCAPE DESIGN

解读青城

项目用地范围　Scope of land of the project

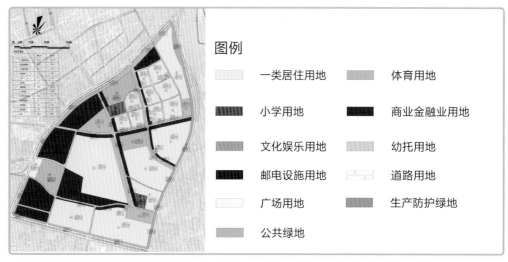

青城世界旅游特别规划区　Special planning area of Qingcheng World Tourism

青城（豪生）国际酒店
区位说明

LOCATION EXPLANATIONS OF QINGCHENG (HOWARD JOHNSON) INTERNATIONAL HOTEL

基地位于青城山——都江堰，国际文化休闲度假旅游片区。该片区以世界双遗产文化为依托，将道教文化和水利文化推向世界舞台，同时将当今世界绿色休闲元素注入本规划区，形成具有国际影响力的度假旅游胜地。

The base is located in Qingcheng Mountain---Dujiangyan Dams, the international cultural leisure holiday tourism region. The region is based on world double cultural heritage and promotes Taoism culture and water conservancy culture to the world. Meanwhile, it adds the green leisure element of today's world to the planning area and has formed a holiday resort with international influence.

意境图

意境图

意境图

青城(豪生)国际酒店景观设计
HOWARD JOHNSON LANDSCAPE DESIGN

酒店设计概念

原始设计总图
Original General Design Layout

初期景观思路
INITIAL LANDSCAPE INSPIRATION

酒店景观方向
LANDSCAPE DIRECTIONS

现代格调：
材料，风格，精致，简练，纯净

地域色彩：
青城山：宗教，自然地貌，民俗色彩
川西：构筑，宅地院落，植被

地景塑造：

地形：浅丘，临盆，曲折的变化。

水涧：需以蜿蜒溪涧，局部错落跌水体现，切忌大规模滥用溪流水系。

植被：树林，竹等植被季象变化；以林衬景，以绿刻画空间，营造现代雅致高档绿景；在庭院植被设计之中，通过景致化的搭配与精致的修剪，使整个景观呈现出档次，同时表现出如传统盆景艺术中的造型艺术。

景石：提炼画龙点睛的石料运用，碎石中具有的中式简洁韵味；整洁精致的"石尚"营造，代表着现代高档次场所的雅致原貌。

Model of Landscape

Terrain: Shallow mound, lying-in, winding change

Water stream:
wriggle creek stream, the water stew at random, do avoid abuse of creek water overly.

Vegetation:
Forest, bamboo etc. vary in seasons; the use of green plants as a space character creates an elegant and modern environment. Scientific arrangement and the pruning of the plants present visitor a tasteful hotel and display the traditional art of bonsai.

Sight of stone:
To purify the use of stone and add the finishing touch to show the simplicity of the Chinese style from the gravel; It is a topquality and elegant place showing from the tidiness and delicacy of "Stone fashion".

Modern Pattern
Material, style, delicacy, pithy, purity

Local Character
Qingcheng Mountain: religion, natural relief, folk-custom
West of Sichuan: building, curtilage courtyard, vegetation

设计原理
PRINCIPLE OF DESIGN

营造人性化的景观环境，体现青城国际酒店特色，营造植物与地形的变化，合理的水景使用，形成"日出青城，月洒酒店"的阴阳互补关系。

Reasonable use of waterscape and combination of plant and terrain to build a humanistic surroundings,"Sunrise in Qingcheng, moonlight pours on the hotel". We strive to form the balance of Yin and Yang and the feature of Qingcheng hotel international.

景观策略意向
QINGCHENG LANDSCAPE STRATEGY

青城天下幽 —— 白鹭林涧恬

青城幽天下 —— 鹭洲隐云海

身隐青城 —— 心安鹭洲

幽隐于青山 —— 息从于鹭洲

青城（豪生）国际酒店景观设计
HOWARD JOHNSON LANDSCAPE DESIGN | 酒店设计概念

青城山语间

幽　you　隐也。从山中，亦聲。　於虬切
恬　tian　安也。从心，甛省聲。　徒兼切

景观主轴

幽山 ⇨ 怡绿 ⇨ 灵水 ⇨ 秀居 ⇨ 一种栖息于此的享受场所

MAIN SIGHT

Quiet Mountain → Pleasant greenery → Pure water → Elegant living → An enjoyable habitat

早期设计总图
Early main design

近期设计总图
Present main design

酒店景观设计说明
EXPLANATIONS OF THE HOTEL LANDSCAPE DESIGN

青城（豪生）国际酒店由成都建筑工程集团总公司投资兴建，世界500强美国温德姆集团旗下的豪生国际酒店集团管理的国际酒店。酒店位于世界自然文化遗产、道教发源地青城山前山风景区，紧邻青城前山新山门。酒店项目总投资人民币14亿元，占地22公顷，周围绿树成荫，环境优雅、空气清新，被喻为天然氧吧，是一座集商务会议、休闲度假、避暑养生为一体的综合型五星级国际酒店。

景观设计上，将酒店的标准景观区、温源谷（温泉区）、高尔夫练习场、道缘食府、生态停车场等，以原生态的设计手法结合现代空间理念，营造出具有当地特色的人文与自然景观效果。同时，以当地的道家文化为基础，将道家哲理体现在景观的环境营造中，体现青城山清幽脱俗的特色。

Qingcheng (Howard Johnson) International Hotel is invested and built by Chengdu Engineering Construction Group, administered by Howard Johnson International Hotel Group which is owned by American Group---one of the Fortune Global 500. Located in the front mountain scenic spot of Qingcheng---a world cultural heritage site and birthplace of Taoism and closely beside the new front gate of Qingcheng front mountain, this hotel covers an area of 22 hectares and boasts a main investment as much as 1.4 billion RMB. There are beautiful environment, fresh air and numerous trees shading off heat for it, causing people to also call it a natural oxygen bar. It is a comprehensive five-star international hotel combining commercial meetings, leisure and holiday relaxing, shade-seeking in summer and health-preservation.

As for landscape design, it combines proto-ecological designing way and modern space concept to create a natural landscape of local cultural characteristics which can be seen through such areas as standard scenic spots of the hotel, Wenyuan valley (hot spring area), golf practicing course, Daoyuan restaurant, ecological parking area and the like. Meanwhile, it is based on the local Taoism culture and embodies philosophy of Taoism within creation of landscape showing Qingcheng mountain's tranquil and refined environment.

Johnson® 国际酒店

青城（豪生）国际酒店景观设计 | 酒店入口景观
HOWARD JOHNSON LANDSCAPE DESIGN

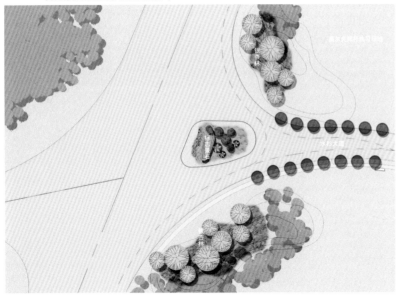
106国道主入口平面图　Plan of the main entrance of 106 National Highway

东软大道主入口　Main entrance of Neteye Avenue

酒店入口景观
ENTRANCE LANDSCAPE

青城（豪生）国际酒店两大入口分别位于106国道的北入口与东软大道的西南入口，入口整体展示上体现大气挺拔的形象，以及局部精致细腻的点缀，形成视觉体量以及触感上的对比。

入口的设计上，采取标志的深山景石与浓密的当地植物组合，体现与青城山相呼应的景观效果。同时也将酒店自身的标志导向物布置于景观中，尽量与周边的景区环境相互协调，相互呼应。

在进入酒店区北面车型道的两侧分别先以水杉，后以银杏为主的林荫道将酒店氛围进行烘托，以静谧的效果隔绝外围车流的繁杂。而且通过时令花卉以及当地植被的搭配，在清幽中展现色彩的鲜活与青城山润泽的印象。

东软大道主入口立面图　Elevation drawing of the main entrance of Neteye Avenue

东软大道主入口效果图 Rendering drawing of the main entrance of Neteye Avenue

There are two main entrances in Qingcheng (Howard Johnson) International Hotel. One is located in the northern entrance of 106 National Highway and the other in the southwest of Neteye Avenue. The whole design of the two entrances reflects a tall and straight image of great momentum. Besides, the delicate and exquisite partial ornaments form comparison of visual dimension and sense of touch.

In the design of the entrances, the symbolic mountain landscape stones are combined with local bushy plants. It reflects the perspective effect of correspondence to Qingcheng Mountain. Meanwhile, the hotel's symbol guide is designed in the landscape to coordinate with the surrounding environment.

Along the northern roadway leading to hotel area, the sequoias, and then, the ginkgos come in sight on both sides. Such a boulevard adds to the atmosphere of the hotel and insulates the outside noise with the tranquil effect. What's more, it demonstrates the impression of fresh color and glossy Qingcheng Mountain with the match of seasonal flowers and local plants.

106国道主入口 —— 水杉大道夜景
The Main entrance of national highway — night scene of Metasequoia Glyptostroboides Avenue

水杉大道
METASEQUOIA MAIN STREET

　　阵列式水杉所带来的大气与规整,让人不由的肃然而生敬意。配合其它大型背景乔木、花灌与草坪,营造干净利落的车行道两旁礼仪式景观,不但衬托了这份庄严,同时添加了些许静谧和安详,突显出酒店原生态品质。

　　The magnificent and neat metasequoia arise people's deep respect. At the same time, there are other large-scaled arbors, bushes and lawns, and they form a tidy and solemn roadway. They also add some peace and natural scenery to the hotel.

水杉大道尽端 —— 交通绿岛
The end of Metasequoia Glyptostroboides Avenue — the Traffic Green Island

自然生态湖泊，感悟青山幽水的惬意。 The natural lakes provide the enjoyment of green hills and blue waters

入口交通分流区 Traffic dispersion zone at the entrance

弧形舞台广场
THE ARCH STAGE SQUARE

　　林地水岸，山泉之畔，掩映着独具风情的亲水广场。拉近人与自然山水的距离，感受白鹭在湖面掠影的惬意画面，标志性的木架白帆与广场台地的结合，为大众营造青城景区绿幽环境中清爽的情趣小空间。

The square beside water is in vicinity to the forest and spring. It narrows the distance between people and nature so that people can enjoy the sight of egrets flying above the water. The white sail and the square have formed cool and appealing space in a peaceful environment.

主入口银杏大道鸟瞰 Metasequoia Glyptostroboides Avenue

高大银杏成排而列，勾勒独特风景线
Tall Ginkgos standing in a row and presenting unique scenery

银杏大道
GINGKO MAIN STREET

银杏树被誉为中国的国树，是世界上最古老的树种之一，有着"活化石"的美称。位于酒店正入口前场的银杏树采用种植池形式阵列排序，如华章备序。深秋时分，层林尽染，灿灿金色望去煞是好看，景致动人。

Gingko is renowned as the national tree of China, and is one of the oldest tree species in the world. It enjoys the honored name of "living fossil". Located at the main entrance of the hotel, the gingkoes lie in orderly lines as notes on the musical staff. When the late autumn comes, all the hills turn golden and shining of the leaves is just fascinating.

秋风吹拂树叶滑落，
漫眼的金黄，
挺拔而华贵，
生动了酒店的春华秋实。

The falling leaves flutter in the
autumn breeze. The golden yellow in the view and the
tall and luxury trees animate the
progression of seasons of the hotel.

青城（豪生）国际酒店景观设计
HOWARD JOHNSON LANDSCAPE DESIGN

酒店标准景观

酒店大堂主入口广场
Square in front of the hotel hall main-entrance

酒店一期标准景观区
STANDARD LANDSCAPE FIRST-STAGE

标准景观区一期的设计上，首先满足对复杂交通上的疏导，将旅客引导至相应的服务片区，同时在道路的行进上将不同的场所逐一展现在眼前。

其次，通过景观合理的规划，将酒店有限的场地分片区利用，在满足一定户外行走休息场地的同时也通过植物的划分，将过渡的场地以视觉展示的方式加以利用。

酒店标准景观区整体的氛围以开敞大气的参与空间为主，揉进局部生态林地与水系的结合，还原野生梅花鹿的生态环境，并将趣味式的游步道与地形，林地，鹿场，溪流等结合，达到酒店标准景观区公共参与性与生态还原性的共生状态。

Firstly, the design of the firststage standard landscape meets the requirement of traffic dispersion. It guides tourists to appropriate service areas and at the same time, presents various places along the road.

Secondly, through reasonable planning, the design utilizes the limited fields by dividing them into different parts. The division has met the need of walking and rest and meanwhile makes the use of transition place in a visual presenting way by the division of plants.

The participating space of the open air occupies a major part in the overall atmosphere of the standard landscape. The landscape mixes the combination of partial ecological woodland with the river system and also restores the ecological environment of wild sika deer. It also combines the walking trails with the terrain, woodland, deer ranches and streams etc. Thus, the standard landscape of the hotel achieves the coexistence of the public participation and the ecological restoration.

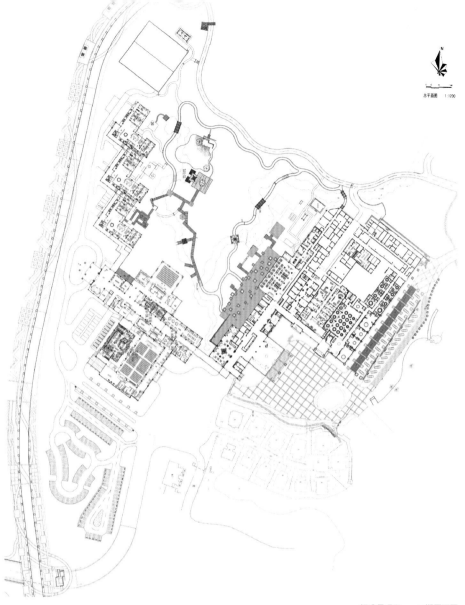

标准景观区一、二期平面图
Plan of the first stage of the standard landscape

大堂入口
LOBBY ENTRANCE

开敞的大尺度空间，尽显不凡气势。以变化的铺装，作为衬托酒店入口形象的前庭广场，同时满足节庆活动的聚集与平时车流系统的缓冲。

The open space shows the uncommon momentum. The front square with the changeable decoration serves as a foil of the hotel entrance image and at the same time, serves as a buffer to satisfy the festival gatherings and the ordinary traffic system.

植物的参差错落给从不同角度观看的每个人不同的感受
The uneven plants with straggly level give people different feelings from different angles

酒店标准景观别墅区
THE VILLA REGION OF THE HOTEL STANDARD LANDSCAPE

远离城市的喧嚣，和着鸟语蝉鸣。吸纳人间仙境飘来的习习清风，细细品味宁静的致远，天人合一的超然境界，充分将别墅建筑与环境，人与自然高度的和谐统一。一切从这里开始……

Far away from cities' noises, accompanied by singing of birds and cicadas and breathing wisps of breeze from the earthly fairyland you are able to peacefully enjoy tranquility and let your imagination fly feeling the supra-realm of harmony between human beings and nature. Here there is excellent harmony and unity between cottages and environment, between human beings and nature. Everything starts from here

林荫下一块休闲木平台，一场悠闲下午茶　　A leisurely platform under the tree-shade, a cup of relaxing afternoon tea.

酒店一期标准景观区
休闲聚会平台
RELAXING AND GATHERING PLATFORM

木平台与树池的结合，营造出静谧、舒适的酒店户外休闲空间。树池将平台分隔成许多相对独立而又彼此联系的小空间；竖向上大小乔木层层叠叠的结合丰富了空间层次，简洁现代的软装布局也与酒店室内相呼应。

The wooden platform and the trees create a peaceful and comfortable outdoor leisure center. The platform is divided into many small independent and also related spaces by the trees. The upright trees in all sizes enrich the layers of the space. The simple and modern decoration is the same style as the hotel interior decoration.

青城（豪生）国际酒店景观设计
HOWARD JOHNSON LANDSCAPE DESIGN

林荫中的休息亭　The rest platform under the trees

浅丘栈道休憩平台　The rest platform on the gentle hill

酒店一期标准景观区
芭蒂苑
BADI PARK

蜿蜒迂回的木栈道上，两旁有溪水和树木的陪伴，朴质的美感浸染心灵。木桩与麻绳制成的护栏，加之高低错落的植物就在附近，使人更加感受到自然原生态的气息。

Accompanied by the stream and trees on both sides, visitors on the winding plank road will be moved by the beauty of simple. The guardrail is made by wooden peg and hemp rope, along with the high and low plant beside, visitor can taste the freshness of nature.

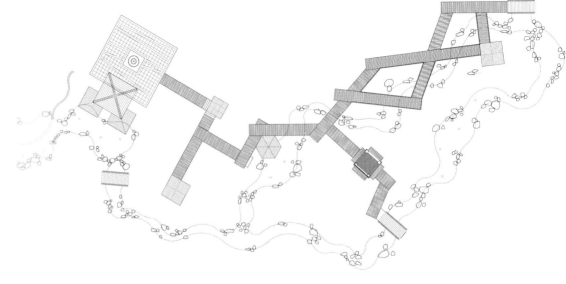

设计初期的栈道平面图　The plank road surface in the original plan

趣味木栈道，原生的景观 Interesting plank roads and primary natural landscape

青城（豪生）国际酒店景观设计
HOWARD JOHNSON LANDSCAPE DESIGN

酒店标准景观

大面积阳光草地带来极佳的视线与舒松的心境　Large scale sunshine grassland brings excellent view and comfortable mood

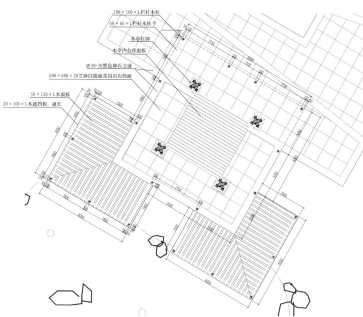

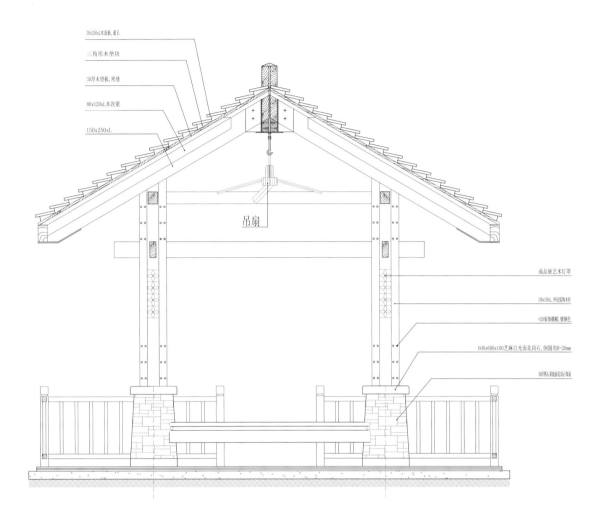

鹿场旁的景观休息亭 The landscape rest pavilion beside the deer ranch

宽阔的休息草坪 Broad rest grassland

酒店一期标准景观区
鹿苑
DEER GARDEN

曲径通鹿舍，古朴近天然。低矮的草屋，木桩制成的栅栏，看似随意放置的石块，以及层层自然式的跌水穿插，这一切都为了让此屋的鹿主人能感受到家的味道。

There is a winding path to the deer house, and the scene is simple and natural. The low grass house, wooden peg fence, randomly laid stones and the waterfall are to make visitors feel at home.

鹿舍一角 A corner of the deer house

古朴野趣的鹿舍 The primitive deer house offering wild delight

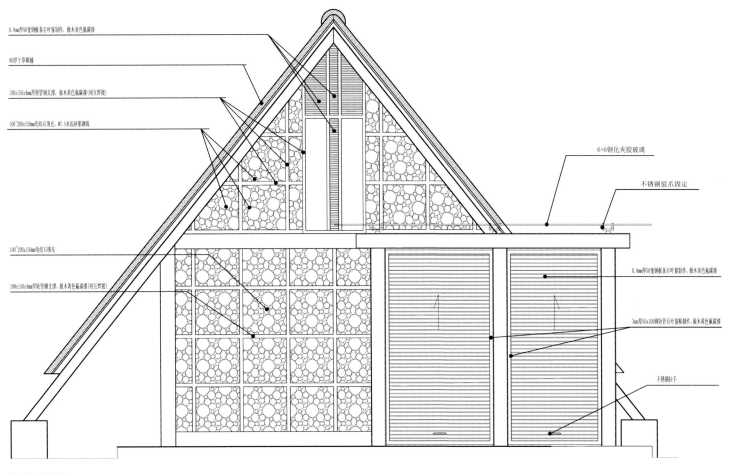

鹿舍立面设计图　The Elevation drawing of deer house

蜿蜒的小径直通向古朴的鹿舍　Winding path leading to the primitive deer house

二期空中景观平台　The second-stage airspace landscape platform

酒店二期标准景观
STANDARD LANDSCAPE SECOND STAGE

　　在二期景观的营造上，延续一期景观的风格，以原生态山林的再生为出发点，强化酒店二期林中居住效果，以大面积的原生林地夹杂着生态步道系统。

　　二期景观将适当的林地与建筑穿插，使旅客在室内也能感受到临青城清幽式的养生气息，彰显青城国际酒店作为山中酒店的独特气质。

The construction style of the second-stage landscape follows the same style in the first stage. It starts with the regeneration of the original mountain forests and strengthens the effect of living in the forest. Besides, the roads are distributed in a large area of original woodland.

The second-stage landscape combines the woodland with the architecture appropriately. It enables the tourists indoors to feel the peaceful freshness of Qingcheng Mountain as if they were in the mountain. The design represents the unique character of Qingcheng International Hotel as a hotel in the mountain.

酒店二期入口绿岛　　The second-stage green island at the entrance of the hotel

酒店主入口灯饰　　Decorative lighting at the main entrance of the hotel

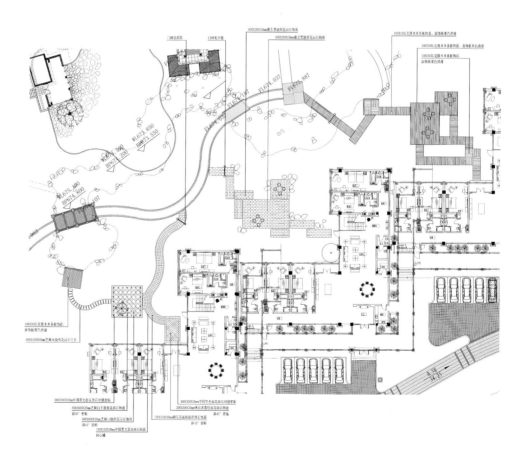

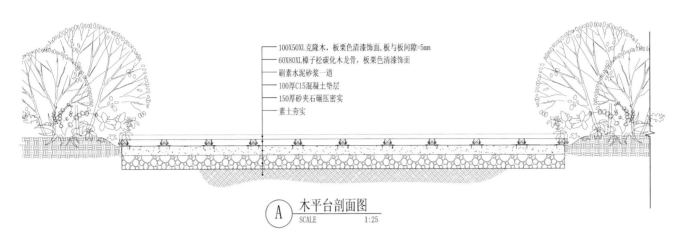

A 木平台剖面图　SCALE 1:25

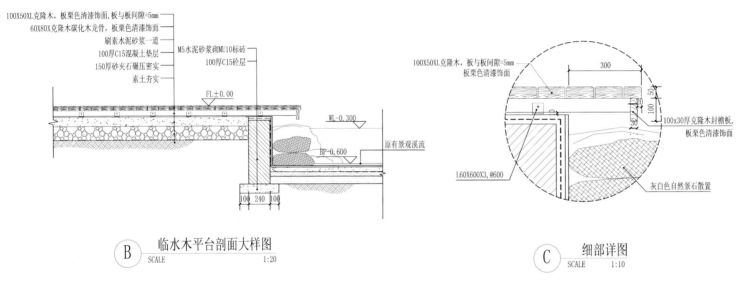

B 临水木平台剖面大样图　SCALE 1:20

C 细部详图　SCALE 1:10

酒店二期VIP客房停车场　The second-stage VIP guest rooms of the hote Parking Lot

VIP客房入口　Entrance to VIP guest rooms

VIP客房和屋顶景观

在沿续酒店整体风格的基础上，

更加精致考究，

用心营造的景观，

无不透露出独特的典雅与华贵。

VIP guest rooms and roof landscape are more delicate and
fastidious on the basis of overall style of the hotel
and the painstaking design of the landscape construction
reveals the unique elegance and luxury.

酒店二期顶屋套房配套空中花园　　The hotel's second-stage roof suite with the hanging garden

酒店生态停车场 Ecological parking lots

酒店生态停车场
ECOTYPE PARKING LOT

将生态停车区与自然植被以及地形结合，将汽车隐匿于浅丘林地之中。生态停车场延续原有地形的起伏变化，梳理原生态杂乱的植被，以有序的场地规划，带来独具匠心的设计。

The design combines the ecotype parking lot and the natural vegetation. The cars are hidden in the low-living trees. The ecotype parking lot takes advantage of the terrain, and change the raw and littery vegetation to an orderly layout so that it shows the elaborate and inventive design.

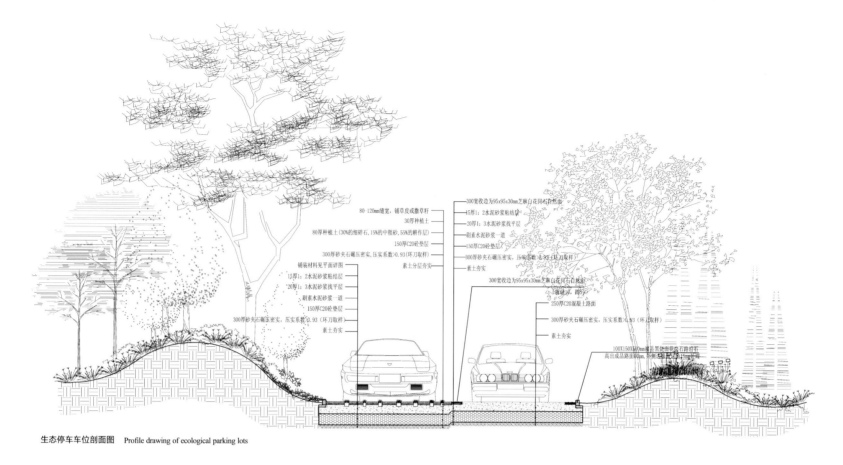

生态停车车位剖面图 Profile drawing of ecological parking lots

生态停车场入口　Entrance to the Ecological Parking Lots

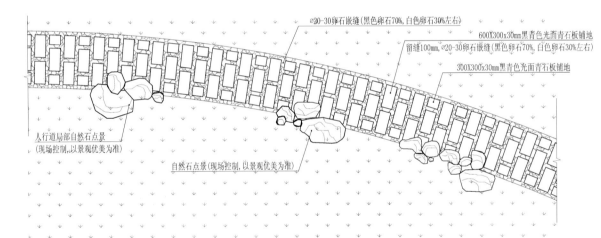

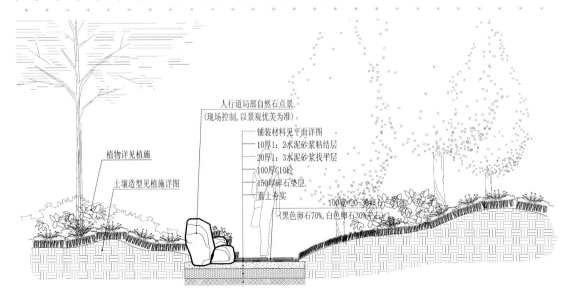

酒店户外停车区　Outdoor parking lots of the hotel

青城（豪生）国际酒店景观设计
HOWARD JOHNSON LANDSCAPE DESIGN

温源谷

分区概念草图
Concept sketches of partitions

初期设计图
Primary design drawing

定案设计图
Final general design drawing

温源谷景观方向
DIRECTION OF WENYUAN LANDSCAPE

塑造宗旨： 不把温泉单纯的看作温泉，寄予其为白鹭洲整体的魅力来考虑

养生健体： 打造地域温泉文化，策划白鹭洲温泉历史

功能选择： 室内池，半室内池，庭院池区，私密汤池，泡泡浴，药池，泡脚池，生物养生池，地热疗养，足底按摩，砂石保健，蒸汽浴，SPA……

休闲放松： 通过多种功能提供消费群体使用与逗留，度假游的配套之一

体验景致： 青城山为大环境，白鹭洲为山下小环境；山景为磅礴幽隐，白鹭洲提供一种精致、秀雅、互动的高档体验

服务享受： 通过提供不同的服务配套设施，以及高档次的室内外营造私密的场所

Aim of creation: Not only take it as a water spring, but also take it into consideration as part of the Heron Island's fascination.

Health and body-building: To build the regional Spring culture, to plot the spring history of Heron Island.

Function selection: indoor pool, half indoor pool, courtyard pool, private pool, bubble bath, Chinese herbal medicine pool, foot wash pool, terrestrial heat pool, foot message pool, stone health care, steam bath, SPA etc.

Leisure and relax: Multiple functional selections for different group of customers; To serve as holiday recreations.

Experience the scenery: Take Qingcheng Mountain as a overall background, Heron Island as a branch environment; the mountain sight is boundless and leisurely. The Heron Island is an experience of delicacy, refined and interactive view.

Enjoyable service: To supply different kinds of services and facilities, high-class atmosphere both indoor and outdoor; to serve as a unique private choice.

意境图　意境图　意境图

温源谷入口景观展示
Landscape stone at the Entrance of weyuan valley

温源谷入口景石　Landscape stone at the entrance of Weyuan valley

温源谷是青城山地区唯一一个星级生态养生温泉场地，把使用者引向户外的青城山境中，将室内外温泉池有机的融合，使得温泉养生池横越室内外的界限，使人犹如在清幽的青城山山林中享受天然温泉水的滋养。

温源谷设计以还原生态，保护生态为出发点，最大限度的将温泉池布置与原生的山林地势相结合，温源谷内外共设27处温泉泡池、10处矿泉水池，以及一组350平米的双温池与30平米水中吧亭，一间溪岸开敞式足疗与地热疗养房，一间桃林的木桶浴房，四间会所式独立温泉浴木房与一处280平米的山涧林荫休息区。

WenYuan Valley is the only Star Ecological Health spa in Qingcheng district. It attracts tourists to Qingcheng Mountain and blends indoor spring and outdoor spring pools perfectly to break the boundaries between them. Thus, people can enjoy the nourishing of natural spring as if they were in the beautiful Qingcheng Mountain.

The design of WenYuan Valley proceeds from restoring and protecting the ecology. The design combines the arrangement of spring pools with the topography of original forest to the greatest extent. There are 27 spring soaking pools indoor and outdoor, 10 mineral water pools, a group of double spring pools of 350 squares meters and pavilions on water of 30 squares meters, an open recovery room of foot massage and ground heating on the lake bank, a budget bathing room among the peach trees, 4 separate spring bathing rooms of club pattern and a rest area of 280 squares meters in the forest.

温源谷建筑气势恢弘，与古木参天；大体量柱头的框架结构使建筑整体与幽静秀美的青城山遥相呼应。
Buildings of Wenyuan valley are of tremendous momentum and as towering as ancient trees; the frame structure of bulky stigma sets the whole building to face the quiet and elegant Qingcheng mountain far ahead.

温源谷所有的功能性场地均完美的与地势结合，并穿插于原生林之间，做到人工养生环境与原始自然环境的有机融合，体现青城山道家养生的哲学价值观。

All the functional places of Wenyuan Valley are designed perfectly in combination with the topography and are scattered in the original forests. The design reflects the organic integration of the artificial health environment with the original natural environment and reveals philosophical values on Taoism health preserving of Qingcheng Mountain.

建筑主入口　Main entrance of the architecture　　标志景石　The symbol stone

室内温泉池　Indoor spring pools

温源谷 — 室内方池
WENYUAN VALLEY
INDOOR POOLS

室内恒温泳池，旁边有配套"小鱼池"以及休息按摩区及吧台。
The constant temperature indoor swimming pool is restful and elegant. There exist a little fish pond and massage zone for relaxation.

室内休息区　Indoor rest area

青城(豪生)国际酒店景观设计
HOWARD JOHNSON LANDSCAPE DESIGN
温源谷

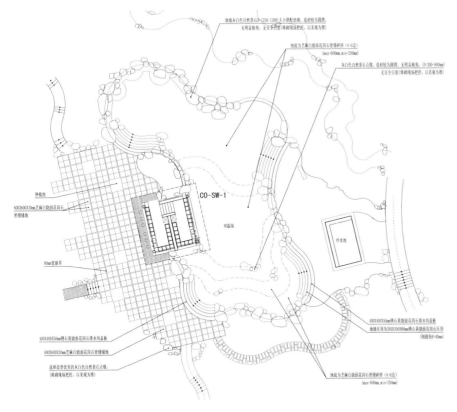

室外双温池平面图
Plan of the outdoor double spring pool

室外双温池休息区　Outdoor rest area of the double spring pool

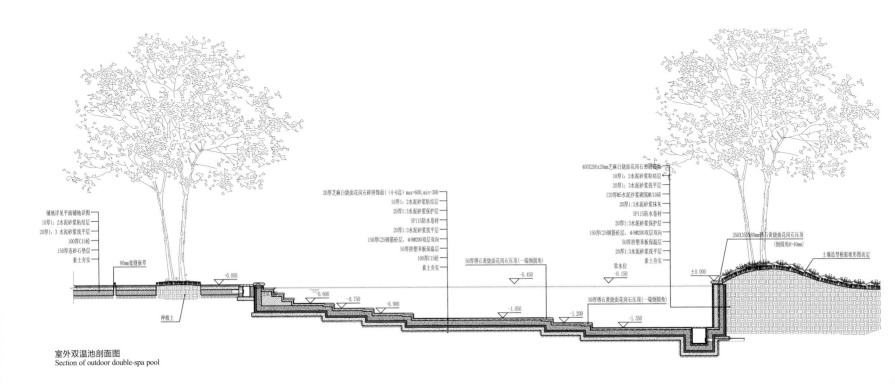

室外双温池剖面图
Section of outdoor double-spa pool

温源谷 — 羲和池区
WENYUAN VALLEY
XI HE POOL

羲和池又名矿泉浴疗，是350平米的双温池。由浅到深的水面，滨水的吧台及躺椅休闲区，聚汇人气，打发时光，或三五成群或独自小憩，定能得其所哉。

This is the 350 square meters double hot spring pool with water from shallow to deep. There is a bar counter and reclining chairs at the water fronts. It's an ideal and restful place to spend your time with friends or on your own.

人性化的叠级池府，满足客人的不同需求，景石与绿树相结合，让您浸泡在山野的自然风光中。
Humanized folding class pool houses combined with landscape stones and green trees meet different needs of different guests and enable you soak in the field of natural scenery.

温源谷 — 阆苑池区
WENYUAN VALLEY
LANG PARK POOL

温泉汤池名碑　The famous monument of the spring bathing pool

阆苑池又名道家药浴，依山而建，错落雅致，返朴求真。与大池不同，它更多的是自然与幽静，隐藏于绿林之中的共享空间。

Langyuan Pool also called Taoist Medicine-bath is built at the foot of the mountain, showing a graceful layout and spirit of returning to purity and simplicity. Different from the big pool, it boasts more nature, tranquility, and sharing space hidden in the woods.

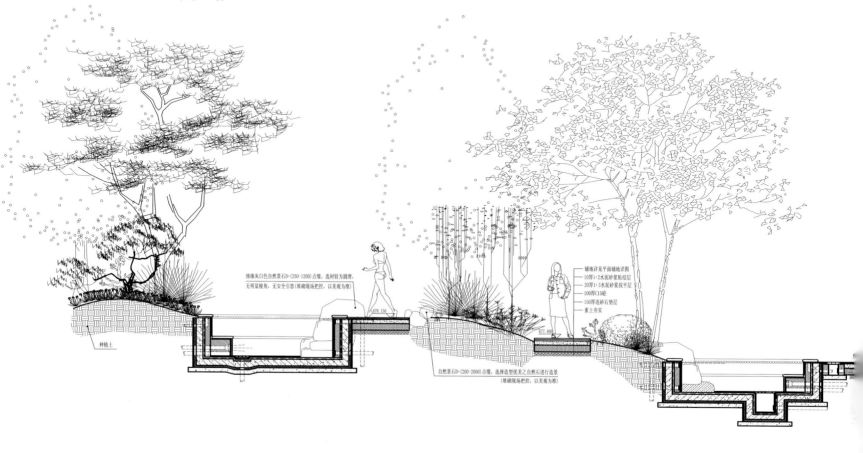

温泉汤池畔休息区　Rest area besides the spring bathing pool

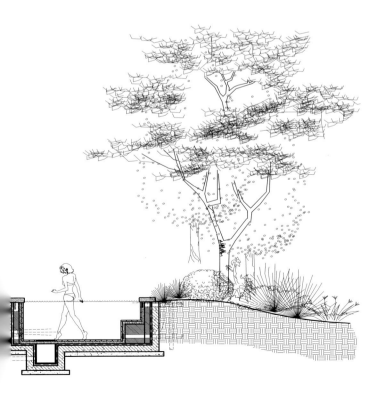

温泉汤池一角　A corner of the spring bathing pool

青城（豪生）国际酒店景观设计 | 温源谷
HOWARD JOHNSON LANDSCAPE DESIGN

藏云地暖木屋入口　Entrance to Cang Yun ground heating cabin

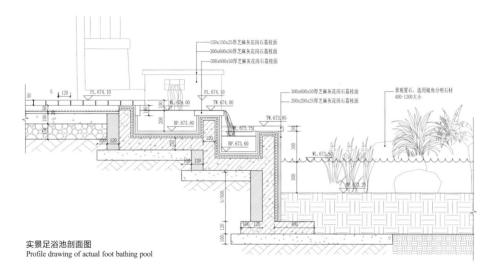

实景足浴池剖面图
Profile drawing of actual foot bathing pool

实景足浴池　Actual scenery of foot bathing pool　　地热石板　Ground heating stone floors

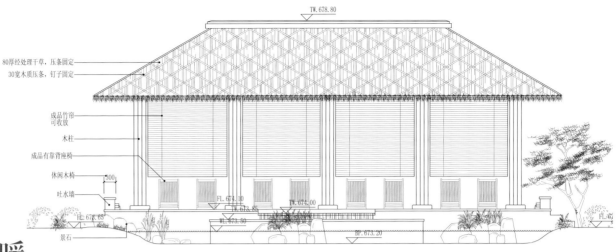

Elevation drawing of Cang Yun ground heating cabin　藏云地暖木屋立面图

温源谷 — 藏云地暖
WENYUAN VALLEY
CANGYUN HEAT SUPPLY

别致的足浴汤池，坐落于水溪山林边。大理石制的地热按摩台，木屋低垂的竹帘，可赏戏水小鱼，可观自然静谧之气息，静尘心之烦杂，无俗务之乱心。

The unique lavipeditum pool is near the forest by the stream side. Enjoying the heat supply massage platform which is made of marble and sitting near the bamboo curtain, you can see and appreciate the little fish and taste the peaceful breath of nature.

专为足浴定制座椅　Customized seats for foot bathing

藏云地暖木屋过道　Corridors of Cang Yun ground heating cabin

青城(豪生)国际酒店景观设计
HOWARD JOHNSON LANDSCAPE DESIGN

温源谷

林间步道 Roads in the forest

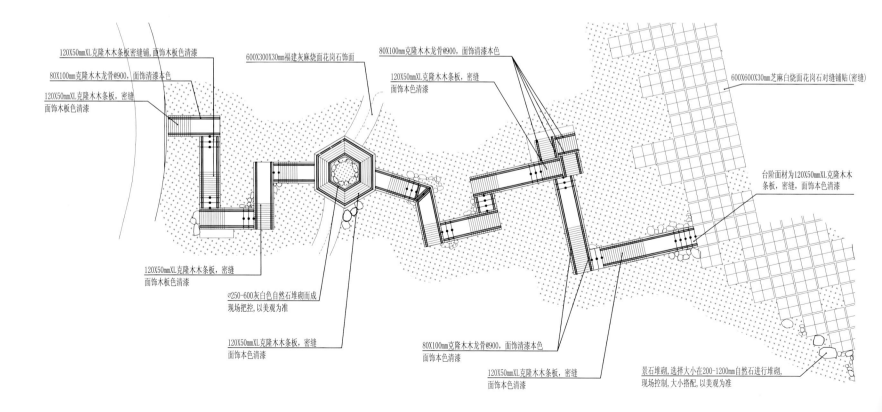

桶浴木屋　Bucket bathing cabin

温源谷 — 道家木桶药浴
WENYUAN VALLEY
TAOIST CASK MEDICINE-BATH

简洁的亭子，低垂的竹帘，大型的木桶，环绕的绿荫，浸泡在这样的意境中，将每个毛孔都打开来，全身的每一个细胞无不愉悦起来。

With the simple pavilion, the hanging bamboo curtain, the broad scale wooden pail, visitors are surrounded by greenness to taste a relaxing joy and it is comfortable enough to open every pore and delight every cell.

私密的林荫围档，别致的山林别墅，让客人对生态温泉的情结以私有的形式释放
Unique mountain forest villas surrounded by private trees enable guests release the emotion for ecological springs in a private way

VIP山林温泉别墅
VIP Villas in the mountain forest

山林别墅汤池一角
A corner of the bathing pool of the Villas in the mountain

温源谷 — VIP山林别墅
WENYUAN VALLEY

VIP VILLA AREA IN THE MOUNTAIN FOREST

　　隐逸于原生林中，独立出来的私享空间，彰显出尊贵与独特，再配合朴实简约的木门及平台，不露声色的彰显出另一种内敛与低调的奢华。

　　Hidden in the pristine forest is the private space to enjoy being distinguished and privacy. The simple style wooden door and the platform, will definitely reveal luxury in a low profile way.

汤池边休息平台　The rest platform beside the bathing pool

山林别墅的指示牌　Signs of the villas in the mountain forest

林地集中休息平台　Rest platform in the forest

温源谷 — 道源仙居
WENYUAN VALLEY
XIANJU TAO PREDESTINATION

以台地休息平台，结合休息树阵以及服务小木房，并与周边林地溪流的惬意环境相融合，营造水疗后在绿氧环境中放松身心的场所。

The Mesa resting floor, combined with tree array and the small wooden service house, along with the stream to form a very cozy environment and it's a restful place after the visitors enjoy the green oxygen hydrotherapy.

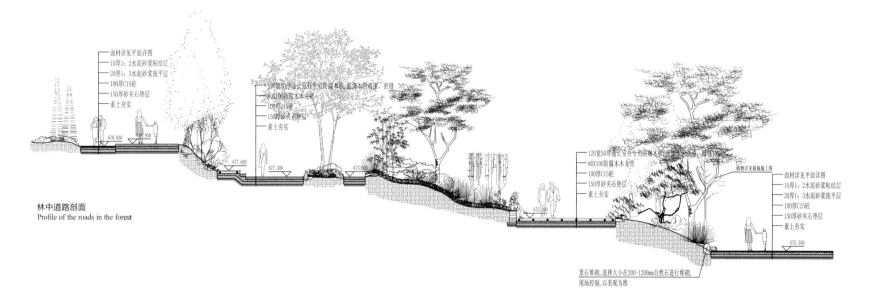

林中道路剖面
Profile of the roads in the forest

青城（豪生）国际酒店景观设计
HOWARD JOHNSON LANDSCAPE DESIGN

道缘休闲岛

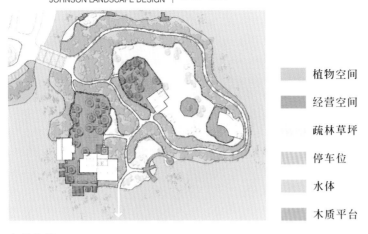

植物空间
经营空间
疏林草坪
停车位
水体
木质平台

空间分析
Spatial Analysis

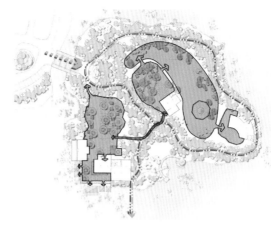

岛上入口
游步道
经营流线

在对岛上游步道的布置上，我们尽量的避免游步道与经营管理流线上的冲突。将两者之间相互的影响减小到最低。

流线分析
Streamline Analysis

景观平面图　Scenery plan

连接道缘岛与酒店的石桥，绿篱围绕，尽显生态野趣
The stone bridge which links Daoyuan Island and the hotel is surrounded by green fences and it reveals the ecological delight

道缘休闲岛 — 入口桥
DAOYUAN LEISURE ISLAND
ENTRANCE BRIDGE

道缘休闲岛为青城国际酒店一处独立的湖中小岛，周边生态极佳，有成群的白鹭栖息于此，此湖也有白鹭湖的美称。

道缘休闲岛清幽宁静，需通过石桥连接上岛，将世间一切繁杂与纷扰杜绝于桥之外。岛上设置养生食府，水岸木房，林夕茶府，以及草地太极八卦广场，各类道家养生理念在岛上详尽展现。

各类功能场所与小岛地势，林地融合，一草一木皆精心维护与布置，整体环境犹如未被打扰的隔世仙岛，使游人近距离感受青城山道家养生的自然和谐理念。

Daoyuan Leisure Island is a separated island in the lake of Qingcheng International Hotel which has a good surrounding ecological environment. Crowds of egrets habitat on the lake everyday and so it is also called the Egret Lake.

Daoyuan Leisure Island is quiet and secluded. It is connected with the out world by stone bridge and all the noises and confusion will disappear after people walk across the bridge. There is the Health Food House on the island, cabins on the lake bank, Lin Xi Tea House and Taiji Ba-gua grass square on the island. Various Taoist health ideas are exhaustively represented on the island.

Various functional places are designed perfectly in combination with the topography and the grassland of the island. All the grass and flowers on the island are carefully protected and well arranged. The whole environment is like a secluded wonderland which enables visitors to comprehend the Taoist health concept of harmony and nature in Qingcheng Mountain.

青城（豪生）国际酒店景观设计
HOWARD JOHNSON LANDSCAPE DESIGN

道缘休闲岛

食府户外用餐区　Outdoor eating area of Food House

道缘休闲岛 — 养生食府
DAOYUAN LEISURE ISLAND
HEALTH-PRESERVATION RESTAURANT

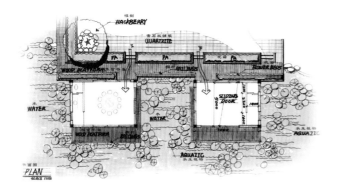

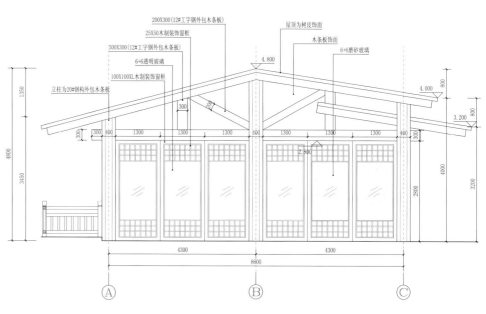

包房木屋立面图　Elevation drawing of private cabins

水岸木屋设计样式 Design of private cabins

包房户外人行木栈道 Wooden plank road for passengers out of the private rooms

水岸木房 Cabins on the lake bank

食府建筑一角 A corner of the Food House architecture

青城（豪生）国际酒店景观设计 | HOWARD JOHNSON LANDSCAPE DESIGN

道缘休闲岛

散布在茶府周边的休闲草地与高大乔木，意味着品茗生活被自然温柔环拥　　Leisure grassland scattered around the Tea House and tall trees mean a life of tasting tea with a tender hug by the nature

道源休闲岛 — 林夕茶府
DAOYUAN LEISURE ISLAND
WOODS-SUNSET TEAHOUSE

茶府休息台地　Tea House rest platform

茶屋 Tea rooms

茶屋入口处 Entrance to Tea House

蜿蜒曲折的木栈道，延伸进别墅…… Winding wooden plank roads extend into the villas…

亲水环道 Water ring road

白露湖边亲水区平面图 Watershed plane along the egret lake

道源休闲岛 — 休闲养生区
DAOYUAN LEISURE ISLAND
RESORT AREA

太极广场上的休闲亭　Relaxing Pavilion on the Taiji Eight-diagram Square

深秋时分，落英缤纷　In the late autumn, petals fall in riotous profusion

太极广场旁的木屋　Special cabins beside the Tai Ji Square

道源休闲岛 — 岛上小品
DAOYUAN LEISURE ISLAND
ACCESSORIAL BUILDINGS ON THE ISLAND

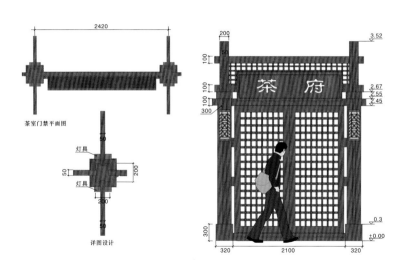

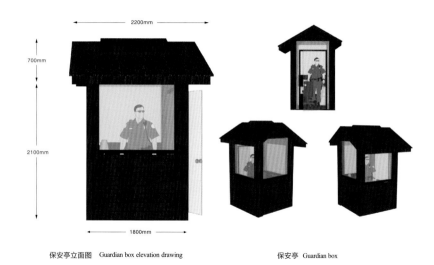

保安亭立面图　Guardian box elevation drawing　　保安亭　Guardian box

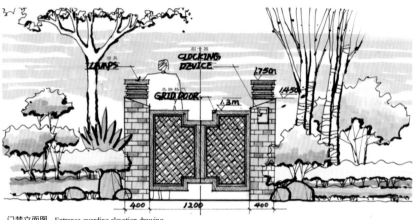

门禁立面图　Entrance-guarding elevation drawing

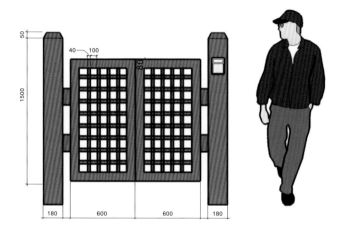

茶社入口牌坊　Memorial archway at the Tea House entrance

食府入口牌坊　Memorial archway at the Food House entrance

门禁系统　Entrance-guarding system

GOLF DRIVING RANGE
高尔夫挥杆练习场

青城（豪生）国际酒店景观设计 | 高尔夫挥杆练习场
HOWARD JOHNSON LANDSCAPE DESIGN

景观设计鸟瞰效果图　The aerial view rendering drawing of the landscape design

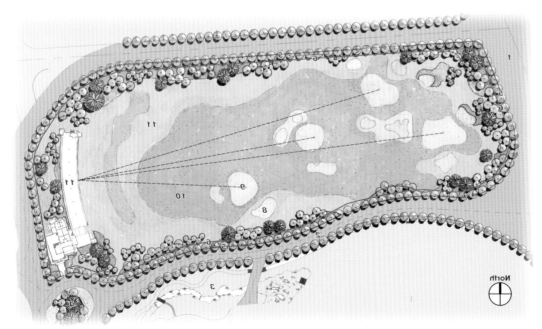

总平面设计图　General plane design

高尔夫
挥杆练习场
GOLF DRIVING RANGE

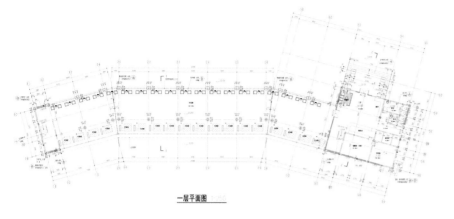

一层平面图

建筑旁的推杆草地　The push-rod grass besides the architecture

挥杆区远眺　Look far into the swing area

作为山脚边唯一一个面向青城山挥球的高尔夫练习场。经过精心选址最终确定于酒店的北面，使球场能最大限度的亲近青城山，并使爱球人士能感受眼前大山的气势之美，面向大山挥出征服的一杆。

整体球场的设计上，利用场地地势的高差，由北向南递减近五米，内部设置连续起伏的浅丘；同时，场地规划设计上将四周地势人为加高，并恢复浓密的生态林地，使整体球场犹如围合在一峡谷之中，只留出通向北面青城山的唯一视线通廊，有效的引导击球人的视线与球路方向。

场地设置一座开放式的钢木结构击球会所，开放式的设计，使击球人士感受三百六十度的景观，并倡导青城山优雅的慢生活理念，使人于此渡过悠闲的时光。

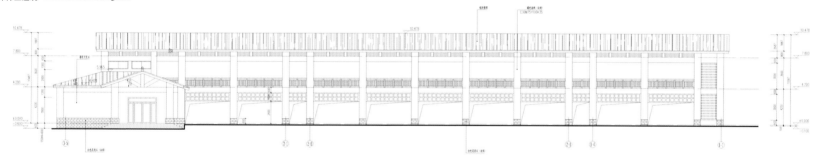

③-⑤-①-① 展开立面图　1:250

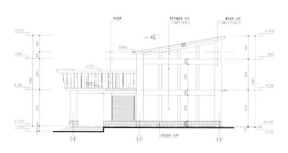

③-⑤-①-Ⓐ 展开立面图　1:250

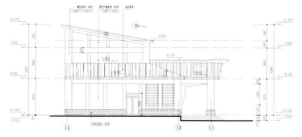

③-Ⓐ-③-⑤ 展开立面图　1:250

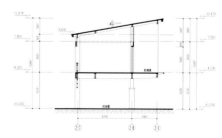

1-1剖面图　1:250

青城（豪生）国际酒店景观设计
HOWARD JOHNSON LANDSCAPE DESIGN

高尔夫挥杆练习场

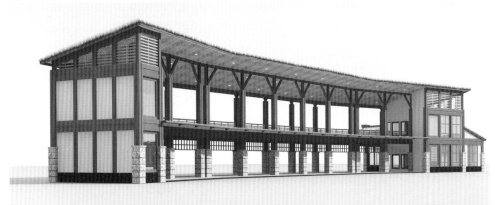

挥杆练习区　Swing practice area

白桦林 The white birch forest

休息区 Rest area

As the only Golf driving range facing Qingcheng Mountain, its location is finally settled in the north of the hotel after careful selection. The location makes the driving range close to Qingcheng Mountain to the greatest extent and enables those who love Golf to enjoy the beauty of the momentum of the mountain in sight. They can swing a conquering golf club in face of the mountain.

The whole design of the driving range takes advantage of the altitude difference of the terrain. Its height constantly decreases about five meters from the north to the south and its inner part has undulating low hillocks. Meanwhile, heightening the surrounding place and restoring the ecological environment make the whole driving range look like being surrounded by a valley. The only visual corridor to the north of Qingcheng Mountain guides the golfer's sight line and ball direction effectively.

In the site, an open ball club of steel structure is set up. The open design; provides golfers with the landscape of 360 angles. The club also advocates the idea of elegant low-pace life in Qingcheng Mountain to make people enjoy leisurely and comfortable days.